WORLD OF VOCABULARY

BLUE

Sidney J. Rauch

Alfred B. Weinstein

Assisted by Muriel Harris

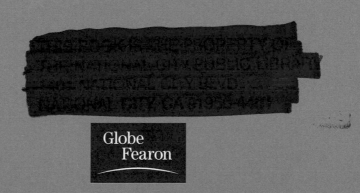
Globe
Fearon

Photo Credits

p. 2: Courtesy of U.S. Space Camp, Huntsville, Alabama; **p. 5:** Courtesy of U.S. Space Camp, Huntsville, Alabama; **p. 7:** Courtesy of U.S. Space Camp, Huntsville, Alabama; **p. 8:** UPI/The Bettmann Archive; **p. 11:** Photofest; **p. 13:** Photofest; **p. 14:** Alice Waters/Chez Panisse; **p. 17:** Alice Waters/Chez Panisse; **p. 19:** Deanne Fitzmaurice/*San Francisco Chronicle*; **p. 20:** Courtesy of HarperCollins Publishers; **p. 23:** George Motz; **p. 25:** George Motz; **p. 26:** Wide World Photos; **p. 29:** Wide World Photos; **p. 31:** Wide World Photos; **p. 32:** UPI/The Bettmann Archive; **p. 35:** UPI/The Bettmann Archive; **p. 37:** UPI/The Bettmann Archive; **p. 38:** Smeal/Galella, Ltd.; **p. 41:** UPI/The Bettmann Archive; **p. 43:** Smeal/Galella, Ltd.; **p. 44:** Neg. no. 334109. Courtesy of Department Library Services American Museum of National History; **p. 47:** François Gohier/Photo Researchers; **p. 49:** Neg. no. 334109. Courtesy of Department Library Services American Museum of National History; **p. 50:** Peter Jordan; **p. 53:** Peter Jordan; **p. 55:** Peter Jordan; **p. 56:** Wide World Photos; **p. 59:** Wide World Photos; **p. 61:** Wide World Photos; **p. 62:** Frederic Reglain/Gamma Liaison; **p. 65:** Frederic Reglain/Gamma Liaison; **p. 67:** Frederic Reglain/Gamma Liaison; **p. 68:** Wally Amos; **p. 71:** Wally Amos; **p. 73:** Wally Amos; **p. 74:** Ron Galella; **p. 77:** Ron Galella; **p. 79:** Photofest; **p. 80:** Wide World Photos; **p. 83:** Wide World Photos; **p. 85:** Wide World Photos; **p. 86:** Wide World Photos; **p. 89:** Wide World Photos; **p. 91:** Wide World Photos; **p. 92:** Leo De Wys; **p. 95:** Leo De Wys; **p. 97:** UPI/The Bettmann Archive; **p. 98:** Wide World Photos; **p. 101:** Wide World Photos; **p. 103:** Wide World Photos; **p. 104:** Wide World Photos; **p. 107:** Todd Webb/Courtesy Amon Carter Museum, Fort Worth, Texas; **p. 109:** Todd Webb/Courtesy Amon Carter Museum, Fort Worth, Texas; **p. 110:** The Bettmann Archive; **p. 113:** The Bettmann Archive; **p. 115:** Wide World Photos; **p. 116:** ICM Artists Limited; **p. 119:** ICM Artists Limited; **p. 121:** ICM Artists Limited.

World of Vocabulary, Blue Level, Third Edition

Sidney J. Rauch • Alfred B. Weinstein

ISBN 0-8359-1298-1

Printed in the United States of America

9 10 11 12 13   06 05 04 03

1-800-321-3106
www.pearsonlearning.com

AUTHORS

Sidney J. Rauch is Professor Emeritus of Reading and Education at Hofstra University in Hempstead, New York. He has been a visiting professor at numerous universities (University of Vermont; Appalachian State University, North Carolina; Queens College, New York; The State University at Albany, New York) and is active as an author, consultant, and evaluator. His publications include three textbooks, thirty workbooks, and over 80 professional articles. His *World of Vocabulary* series has sold over two and one-half million copies.

Dr. Rauch has served as consultant and/or evaluator for over thirty school districts in New York, Connecticut, Florida, North Carolina, South Carolina, and the U.S. Virgin Islands. His awards include "Reading Educator of the Year" from the New York State Reading Association (1985); "Outstanding Educator Award" presented by the Colby College Alumni Association (1990); and the College Reading Association Award for "Outstanding Contributions to the Field of Reading" (1991). The *Journal of Reading Education* selected Dr. Rauch's article, "The Balancing Effect Continue: Whole Language Faces Reality" for its "Outstanding Article Award," 1993-1994.

Two of the *Barnaby Brown* books, The Visitor from Outer Space, and *The Return of B.B.* were selected as "Children's Choices" winners for 1991 in a poll conducted by the New York State Reading Association.

Alfred B. Weinstein is the former principal of Myra S. Barnes Intermediate School (Staten Island, N.Y.). Dr. Weinstein has taught extensively at the secondary school level, and he has served as an elementary school principal and assistant principal. He has been a reading clinician and instructor at Hofstra University Reading Center. At Queens College he gave courses in reading improvement, and at Brooklyn College he taught reading for the New York City Board of Education's in-service teacher training program. He was head of Unit 1 of the Board of Examiners and supervised the licensing of teachers, supervisors, administrators, psychologists, and social workers for the New York City Board of Education. He is vice-president of the Council of Supervisors and Administrators of Local 1 of the AFL-CIO. Dr. Weinstein has been listed in *Who's Who in the East* since 1982.

Dr. Weinstein is a contributor to the Handbook for the Volunteer Tutor and one of the authors of Achieving Reading Skills. With Dr. Rauch, he is coauthor of *Mastering Reading Skills*.

CONTENTS

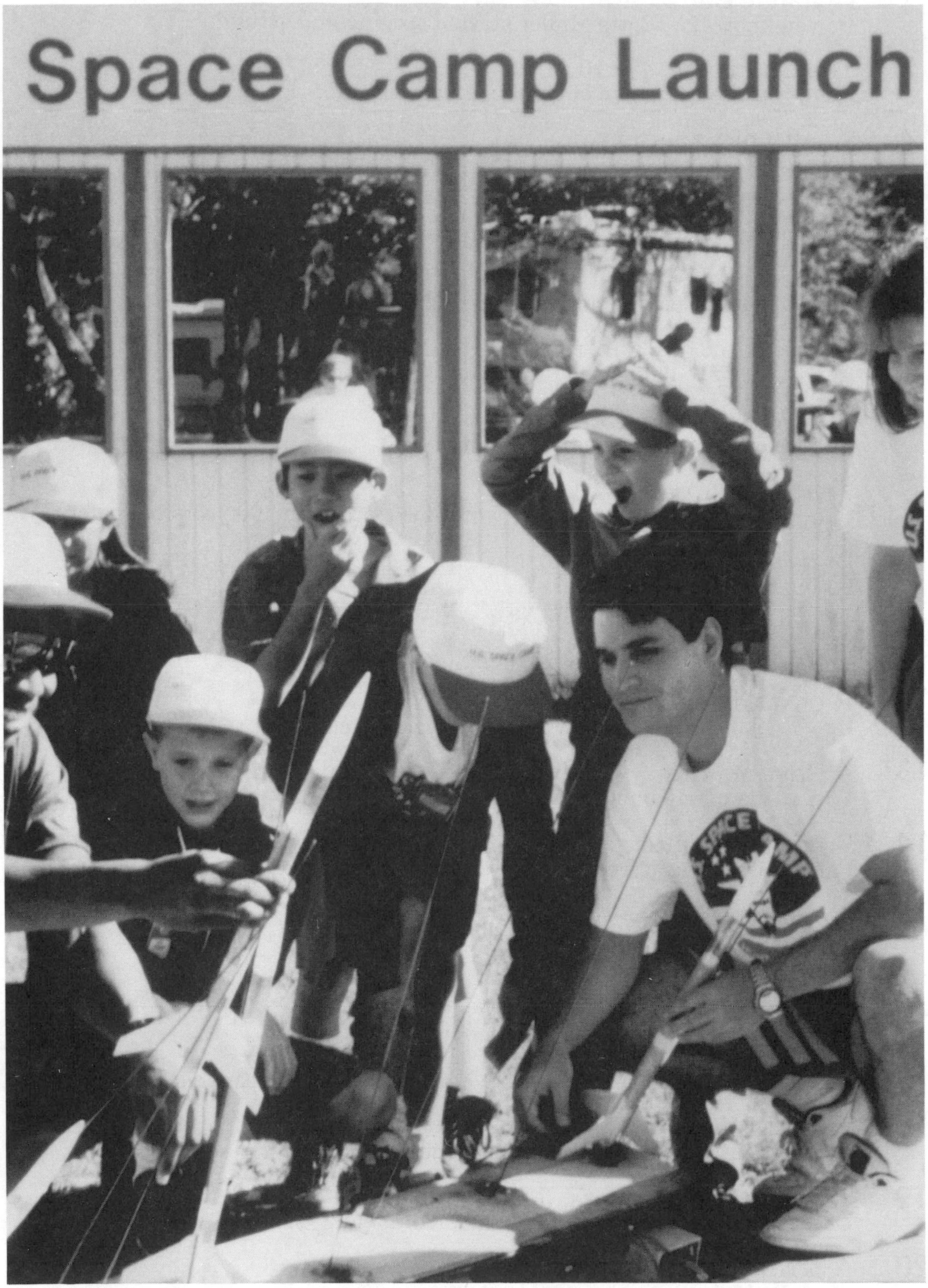
Space Camp Launch

1 AN UNUSUAL CAMP

Do you feel *restless* here on Earth? Would you rather be flying in a space shuttle? If that sounds exciting, consider Space Camp.

In 1982, about 800 young people enrolled in the first Space Camp, located beside NASA's Marshall Space Flight Center in Huntsville, Alabama. In 1988, a second Space Camp opened near the Kennedy Space Center in Florida. Even then few people could have *predicted* the great success of this special camp.

Now 25,000 campers a year experience *intensive* training similar to what *actual* astronauts receive. Campers bounce off walls in nearly zero gravity and try out an apparatus that gives them the feeling of floating in outer space. Between these activity sessions, campers taste space food and learn about space travel from the camp *instructors.*

At the end of the week, one team of campers sits in an imitation Mission Control and directs the launch of a model space shuttle. Another team rides the model as it is launched and orbits Earth. The teams face frightening *circumstances* set up by their instructors. Some teams might have to deal with an *abrupt* loss of oxygen or engine failure, while others might have to avoid a *probable* crash with a meteor. The teams must not let their attention *falter.* They have to work within a slim *margin* of error to land the shuttle safely.

The idea behind Space Camp is to encourage young people's interest in space. It must be working because some of the first graduates of Space Camp are now in military academies, preparing to become astronauts. They plan to board real space shuttles someday soon. At Space Camp, they launched not only a shuttle but their careers.

UNDERSTANDING THE STORY

>>>> *Circle the letter next to each correct statement.*

1. The statement that best expresses the main idea of this selection is that
   a. space travel can be frightening and dangerous.
   b. Space Camp helps young people learn more about flying space shuttles.
   c. Space Camp allows young people to experience space travel firsthand.

2. From this story, you can conclude that
   a. Space Camp strengthens young people's problem-solving skills.
   b. most space campers intend to become astronauts.
   c. astronauts who have attended Space Camp can skip some of the training they would usually receive.

MAKE AN ALPHABETICAL LIST

>>>> *Here are the ten vocabulary words in this lesson. Write them in alphabetical order in the spaces below.*

margin	instructors	predicted	circumstances	restless
abrupt	probable	intensive	actual	falter

1. _____   6. _____

2. _____   7. _____

3. _____   8. _____

4. _____   9. _____

5. _____   10. _____

WHAT DO THE WORDS MEAN?

>>>> *Following are some meanings, or definitions, for the ten vocabulary words in this lesson. Write the words next to their definitions.*

1. _____ uneasy; bored

2. _____ described what would happen in the future; forecasted

3. _____ real

4. _____ likely to happen

5. _____ concentrated

6. _____ to hesitate; to fail or weaken

7. _____ sudden

8. _____ teachers; leaders

9. _____ conditions

10. _____ a border; the space allowed for something

4

FIND THE ANALOGIES

>>>> An **analogy** is a relationship between words. Here's one kind of analogy: *raindrop* is to *wet* as *sunlight* is to *hot*. In this relationship, the first word in each pair is an object and the second word in each pair describes the object.

>>>> *See if you can complete the following analogies. Circle the correct word or words.*

1. **Stop** is to **abrupt** as **camper** is to
   a. sudden   **b.** excited   **c.** predicted   **d.** intensive

2. **Training** is to **intensive** as **launch** is to
   a. actual   **b.** restless   **c.** complex   **d.** absorb

3. **Instructor** is to **patient** as **student** is to
   a. plural   **b.** medical   **c.** married   **d.** fascinated

4. **Death** is to **certain** as **marriage** is to
   a. probable   **b.** faltering   **c.** anxious   **d.** lonely

5. **Wages** is to **calculated** as **weather** is to
   a. autumn   **b.** predicted   **c.** vacation   **d.** winter

USE YOUR OWN WORDS

>>>> *Look at the picture. What words come into your mind other than the ten vocabulary words used in this lesson? Write them on the lines below. To help you get started, here are two good words:*

1. _____ realistic _____
2. _____ exciting _____
3. _____
4. _____
5. _____
6. _____
7. _____
8. _____
9. _____
10. _____

5

FIND THE ADJECTIVES

>>>> An **adjective** is a word that describes a person, place, or thing. The adjectives in the following sentences are underlined: The <u>excited</u> campers cheered (person). They slept in a <u>new</u> dormitory (place). The <u>enormous</u> engines began to roar (thing).

>>>> *Underline the adjectives in the following sentences.*

1. Teams must prepare for abrupt changes in the weather.

2. The leader had a suggestion for his restless team.

3. Good instructors are important for a successful camping experience.

4. The team members knew they had a narrow margin of safety.

5. Intensive training paid off as the huge shuttle was launched.

COMPLETE THE STORY

>>>> Here are the ten vocabulary words for this lesson:

margin	restless	abrupt	predicted	instructors
probable	circumstances	actual	falter	intensive

>>>> *There are six blank spaces in the story below. Four vocabulary words have already been used in the story. They are underlined. Use the other six words to fill in the blanks.*

Space Camp _____ give the campers _____ training before they fly the space shuttle. The campers are warned not to get _____ or let their attention <u>falter</u> for a second. The campers discuss with the instructors the <u>probable</u> obstacles they may face and how to deal with them.

Even though campers know they are pretending, the equipment and computer screens make the <u>circumstances</u> seem real. They prepare for an _____ shuttle launch and flight. They know they have a slim <u>margin</u> for error. They realize that not all problems can be _____ ahead of time. The teams understand that mistakes can lead to an _____ crash or explosion of the shuttle.

Learn More About Space Travel

>>>> *On a separate sheet of paper or in your notebook or journal, complete one or more of the activities below.*

Learning Across the Curriculum

Research how living in zero gravity can affect human muscles, especially the heart. Write a short report describing the problems involved and actions astronauts might take to avoid these problems.

Learning Across the Curriculum

Find out how cuts in the federal budget have affected plans for building a space station or revisiting the moon. Then write a proposal suggesting either how to raise more money or what parts of the space program the government should focus on with limited funds.

Broadening Your Understanding

Invite someone who has been to Space Camp to talk with your class about his or her experiences. Or watch a videotape of the movie *Space Camp*. Discuss which parts of the movie plot you think could come true and which parts seemed to be fiction.

2 DYNAMIC BETTE

The young woman stood before the curtain in the Palace Theater in New York City, her head bowed. The $\boxed{frantic}$ applause of the audience brought tears to her eyes. Bette Midler's $\boxed{debut}$ at the Palace Theater was a tremendous success.

This setting was very different from the gentle, rolling hills of Aiea, a small town near the city of Honolulu, Hawaii, where Bette Midler was born. Midler's $\boxed{aptitude}$ for music and acting led her into her profession—show business. She came to the mainland and became a member of an acting group that performed for children. Later, in order to support herself, Midler worked in the coatroom of a nightclub. One day a friend suggested that Midler $\boxed{apply}$ for a role in a play. The producers liked Midler and later gave her an important role in the musical *Fiddler on the Roof.*

Although she started her career as an actress, Midler first became famous as a pop singer. She received $\boxed{ovations}$ for her $\boxed{dynamic}$ stage presence and her $\boxed{renditions}$ of both old and new songs. Later she decided to return to acting. In *The Rose,* a film about a rock singer, Midler firmly established her credentials as a $\boxed{skillful}$ actress. From there, she progressed to films, such as *Ruthless People, Outrageous Fortune, Beaches,* and *For the Boys.* She also had a starring role in the television movie *Gypsy.* Today she is one of the most popular actresses in Hollywood.

Critics $\boxed{commend}$ her performances on the stage and screen. Fans the world over $\boxed{clamor}$ for more from Bette Midler.

UNDERSTANDING THE STORY

 Circle the letter next to each correct statement.

1. The main idea of this story is to
   a. describe the place where Bette Midler was born.
   b. tell the reader about some of the events in Midler's career.
   c. encourage young singers to imitate Midler's style.

2. From this story, you can conclude that
   a. Bette Midler was born into show business.
   b. Midler never really liked acting on the stage.
   c. luck, talent, and hard work have all contributed to Midler's success.

MAKE AN ALPHABETICAL LIST

>>>> *Here are the ten vocabulary words in this lesson. Write them in alphabetical order in the spaces below.*

frantic	renditions	clamor	commend	dynamic
ovations	aptitude	debut	skillful	apply

1. _____   6. _____

2. _____   7. _____

3. _____   8. _____

4. _____   9. _____

5. _____   10. _____

WHAT DO THE WORDS MEAN?

>>>> *Following are some meanings, or definitions, for the ten vocabulary words in this lesson. Write the words next to their definitions.*

1. _____ wild with excitement; out of control

2. _____ a first appearance before the public

3. _____ a natural ability or capacity; a talent

4. _____ full of energy; vigorous

5. _____ to seek a job; to ask for work

6. _____ performances or interpretations

7. _____ having ability gained by practice or knowledge; expert

8. _____ bursts of loud clapping or cheering; waves of applause

9. _____ to praise; to acclaim as worthy of notice

10. _____ to demand noisily; to call for loudly

COMPLETE THE SENTENCES

>>>> *Use the vocabulary words in this lesson to complete the following sentences. Use each word only once.*

dynamic	aptitude	ovations	renditions	apply
frantic	skillful	commend	debut	clamor

1. Standing _____ are the dream of many young entertainers.

2. Midler's audiences grew to expect _____ performances.

3. Midler's _____ of songs always thrill her audiences.

4. Before Midler made her _____, she worked small jobs.

5. Many people _____ for each part in a play, but only a few are chosen.

6. There is a _____ search for a replacement when the star of a show gets sick.

7. Bette Midler has a rare _____ for both comedy and song.

8. No one expected the critics to _____ Midler's performance in the movie *The Rose,* but they all gave it rave reviews.

9. Midler is _____ at relaxing her audience.

10. Fans stand and _____ for more at the end of Midler's performances.

USE YOUR OWN WORDS

>>>> *Look at the picture. What words come into your mind other than the ten vocabulary words used in this lesson? Write them on the lines below. To help you get started, here are two good words:*

1. _____ costumes _____
2. _____ bunny _____
3. _____
4. _____
5. _____
6. _____
7. _____
8. _____
9. _____
10. _____

FIND THE SUBJECTS AND PREDICATES

>>>> The **subject** of a sentence names the person, place, or thing that is spoken about. The **predicate** of a sentence is what is said about the subject. For example:

> The small boy went to the football game.

The small boy is the subject (the person the sentence is talking about). *Went to the football game* is the predicate of the sentence (because it tells what the small boy did).

>>>> *In the following sentences, draw one line under the subject of the sentence and two lines under the predicate of the sentence.*

1. The young woman stood before the curtain.

2. Bette Midler was born in Aiea, Hawaii.

3. Midler became famous as a singer.

4. Critics commended all Midler's performances.

5. Midler performs on the stage and screen.

COMPLETE THE STORY

>>>> Here are the ten vocabulary words for this lesson:

frantic	apply	debut	clamor	aptitude
skillful	dynamic	commend	renditions	ovations

>>>> *There are six blank spaces in the story below. Four vocabulary words have already been used in the story. They are underlined. Use the other six words to fill in the blanks.*

In order to become a <u>dynamic</u> performer, one must first have the _____ necessary to be successful. One must be able to give good _____ of many types of songs. A _____ use of talent and an ability to <u>apply</u> oneself to rigid schedules help to bring fame. <u>Frantic</u> applause of the audience can greet a _____ on the stage. Standing _____ will be given to these artists at every concert.

Fans will always _____ for more music from their favorite artists, and critics will <u>commend</u> notable performances.

Learn More About Popular Singers

>>>> *On a separate sheet of paper or in your notebook or journal, complete one or more of the activities below.*

Learning Across the Curriculum

All cultures have a celebrity like a Bette Midler—a singer who is larger than life. Think about a singer who performs in another language. Listen to one of his or her recordings and to a recording by Bette Midler. Compare the two singers. Would each be popular in the other one's country? Why or why not?

Broadening Your Understanding

Bette Midler's talent as an actress and a singer make her a natural for musicals like *Fiddler on the Roof*. Watch a musical on video. When you have finished, write a review describing your opinion of the performance. Share your report with the class.

Extending your Reading

Read one of these biographies of popular singers, or find a biography of a singer whom you want to learn more about. Now imagine you are writing a movie about this singer's life. Use the events in the singer's life to create the outline for a movie.

Elvis Presley: The King, by Katherine Krohn
Paula Abdul: Straight Up, by Ford Thomas
Gloria Estefan, by Rebecca Stefoff
Whitney Houston, by Keith Elli Greenberg

3 THE CHALLENGE

The *challenge* of climbing a mountain has attracted brave people. Many have failed. Some have been forced to give up. Others have fallen to their deaths or *perished* in severe blizzards.

Climbing Mount Everest has always been the mountaineer's dream. Towering 29,028 feet above sea level, the *summit* of Mount Everest stands higher than any other. Located in the Himalaya Mountains, this icy peak is one of the greatest challenges on Earth.

Edmund Hillary was determined to *scale* this mountain. With his native guide Tenzing, he called together a group of *hardy* mountaineers. Top physical condition was needed for this *perilous* journey.

Hillary led the way with his guide and climbing companion. For a while, the *expedition* went smoothly and without any problems. He warned the other mountaineers about these hidden *chasms.*

Suddenly a snow ledge gave way, and Hillary fell. Tenzing slammed his pickax into the face of the ice-covered mountain, grabbed the life rope, dug in his heels, and held on. The falling man slowly came to a stop on the *sheer* face of the cliff. He was saved by a thin rope and a brave friend.

For a moment, Hillary had come close to death. But the hardy band of mountaineers continued their climb. Finally, they *attained* their goal. As the first to reach the summit of Mount Everest, they had made the mountaineer's dream come true.

UNDERSTANDING THE STORY

>>>> *Circle the letter next to each correct statement.*

1. When asked why he wanted to climb Mount Everest, a famous mountaineer replied, "Because it is there." He meant that
   a. there wasn't a mountain he couldn't climb.
   b. the challenge of conquering a mountain was reason enough.
   c. even if he failed, the mountain would still be there.

2. For a person to climb a mountain, such as Mount Everest, successfully, the climber must
   a. have great confidence in the bravery and skill of the other climbers on the expedition.
   b. spend millions of dollars on equipment and supplies.
   c. study in great detail the maps used by Edmund Hillary.

MAKE AN ALPHABETICAL LIST

>>>> *Here are the ten vocabulary words in this lesson. Write them in alphabetical order in the spaces below.*

challenge	summit	perished	scale	hardy
attained	expedition	perilous	chasms	sheer

1. _____   6. _____

2. _____   7. _____

3. _____   8. _____

4. _____   9. _____

5. _____   10. _____

WHAT DO THE WORDS MEAN?

>>>> *Following are some meanings, or definitions, for the ten vocabulary words in this lesson. Write the words next to their definitions.*

1. _____ a group of people undertaking a special journey, such as mountain climbing

2. _____ steep; straight up and down

3. _____ deep openings or cracks

4. _____ reached; achieved

5. _____ died, usually in a violent manner

6. _____ the peak; the highest point

7. _____ a call to a contest or battle

8. _____ dangerous; hazardous

9. _____ able to take hard physical treatment; bold; daring

10. _____ to climb

COMPLETE THE SENTENCES

>>>> *Use the vocabulary words in this lesson to complete the following sentences. Use each word only once.*

hardy	scale	perilous	challenge	chasms
expedition	summit	perished	sheer	attained

1. Before the _____ could leave, years of planning were necessary.

2. Some people cannot resist the _____ of climbing a huge mountain.

3. The _____ face of the cliff, slowed the mountaineers' progress.

4. The lead climbers had a good chance to _____ the cliff.

5. Only a very _____ person can stand the freezing temperatures.

6. Because the climb was _____, the climbers took safety precautions.

7. They had to watch out for _____ in the mountain walls.

8. Only three people reached the _____, although the others almost did.

9. The group had _____ its goal, but at the terrible cost of two lives.

10. In memory of the climbers who had _____, they erected a stone marker.

USE YOUR OWN WORDS

>>>> *Look at the picture. What words come into your mind other than the ten vocabulary words used in this lesson? Write them on the lines below. To help you get started, here are two good words:*

1. _____ steep _____
2. _____ snow _____
3. _____
4. _____
5. _____
6. _____
7. _____
8. _____
9. _____
10. _____

DESCRIBE THE NOUNS

Two of the vocabulary words, summit and expedition, are nouns. List as many words as you can that describe or tell something about the words summit and expedition. You can work on this with your classmates. Listed below are some words to help you get started.

summit
1. high
2. cold
3.
4.
5.
6.
7.
8.

expedition
1. large
2. careful
3.
4.
5.
6.
7.
8.

COMPLETE THE STORY

>>>> Here are the ten vocabulary words for this lesson:

challenge	summit	perished	scale	hardy
attained	expedition	perilous	chasms	sheer

>>>> *There are six blank spaces in the story below. Four vocabulary words have already been used in the story. They are underlined. Use the other six words to fill in the blanks.*

There are many high mountain peaks in the world that offer a great _____ to mountaineers. For centuries, men and women have tried to _____ their heights. The _____ leader of an expedition must gather together the best mountaineers and equipment. They must plan every step of the way in order to achieve their goal. Despite their great skill and bravery, many mountaineers have perished. Some have slipped on the _____ face of a cliff and fallen into deep chasms. Others have died in blizzards on their _____ journey. When an expedition reaches the summit, there is great joy. The climbers have _____ their goal of conquering a mountain.

Learn More About the Himalayas

>>>> *On a separate sheet of paper or in your notebook or journal, complete one or more of the activities below.*

Learning Across the Curriculum

The Himalaya Mountains, which include Mount Everest, are one of the most impressive mountain ranges in the world. Find out how this mountain range was formed. Then draw a series of illustrations that explain the process that created the Himalayas.

Broadening Your Understanding

Hillary's native guide Tenzing was a Sherpa. Without Tenzing, Hillary would not have been able to reach the summit of Everest. Find out more about these people of the Himalayas. Write what you find about them. What do they eat? What language do they speak? How do they survive in the harsh land of the Himalayas?

Extending Your Reading

Big Foot, the Abominable Snowman or yeti, is said to come from the Himalayas. Read the information on this monster in one of the books below. Then write about whether you believe the Abominable Snowman really exists.

Big Foot, by Ruth Shannon Odor
The Abominable Snowman, by Barbara Antonopolos
Stranger Than Fiction: Monsters, by Melvin Berger

4 WEALTH OF EXPERIENCE

When Walter Dean Myers was growing up in New York's Harlem, little children sang and held hands on their way to Sunday school. As these happy children became teenagers, however, they also became discouraged by the **barriers** facing them. Myers himself **flirted** with gangs and drugs, but he was rescued by his love of reading and writing.

Myers was born in 1937 and was **informally** adopted by Florence and Herbert Dean when he was 3. They taught him to read when he was a **mere** 4 years old. Despite his ability, school was **misery.** Myers lashed out at classmates who teased him because of a speech problem. Soon officials were trying to suspend the angry and **sullen** boy from school—permanently.

Myers **dealt** with rejection by staying home and reading. He was absent from school so much that he showed up one day without realizing that summer vacation had begun! Then he started to hang out on the streets. Partly to escape a death threat from a gang, Myers joined the Army in 1954 on his 17th birthday.

It was not until 1977 that Myers became a full-time writer. He has helped to **define** life in today's cities through more than a dozen **novels.** The young people in his stories face despair and struggle for ways to survive. Through his characters, Myers explores friendship, responsibility, and self-discovery.

His many awards include the 1988 Newbery Honor Book Award for the **publication** of *Scorpions*. Myers's strong dialogue and his combination of humor and hope make his books very popular with young people.

UNDERSTANDING THE STORY

>>>> *Circle the letter next to each correct statement.*

1. The statement that best expresses the main idea of this selection is that
   a. Myers has used his own experiences to write stories that seem very real.
   b. as a teenager, Myers narrowly escaped becoming involved with gangs.
   c. Myers's speech problems kept him from doing well at school.

2. From this story, you can conclude that
   a. Walter Dean Myers no longer lives in Harlem.
   b. Myers would tell beginning writers to focus on what they are familiar with.
   c. Myers always wanted to be a writer.

MAKE AN ALPHABETICAL LIST

>>>> *Here are the ten vocabulary words in this lesson. Write them in alphabetical order in the spaces below.*

publication	flirted	dealt	define	novels
informally	misery	mere	sullen	barriers

1. _____   6. _____

2. _____   7. _____

3. _____   8. _____

4. _____   9. _____

5. _____   10. _____

WHAT DO THE WORDS MEAN?

>>>> *Following are some meanings, or definitions, for the ten vocabulary words in this lesson. Write the words next to their definitions.*

1. _____ long stories about imaginary people and events

2. _____ gloomy; resentful

3. _____ obstacles; walls

4. _____ handled; managed; faced

5. _____ a way of doing something that does not follow exact rules or procedures; casually

6. _____ showed an interest

7. _____ suffering; distress

8. _____ to explain the meaning of

9. _____ only; barely

10. _____ the production of written material into printed form

FIND THE ANALOGIES

>>>> An **analogy** is a relationship between words. Here's one kind of analogy: *barber* is to *haircut* as *farmer* is to *corn*. In this relationship, the first word in each pair is a worker, and the second word in each pair is the worker's product.

>>>> *See if you can complete the following analogies. Circle the correct word or words.*

1. **Police officer** is to **safety** as **author** is to

  a. handcuffs   **b.** paper   **c.** novels   **d.** computers

2. **Teacher** is to **learning** as **construction worker** is to

  a. a building   **b.** a bulldozer   **c.** cement   **d.** misery

3. **Dentist** is to **healthy teeth** as **doctor** is to

  a. injections   **b.** good health   **c.** medicine   **d.** appointment

4. **Salesperson** is to **sale** as **typist** is to

  a. office   **b.** letters   **c.** barriers   **d.** define

5. **Truck driver** is to **transporting goods** as **plumber** is to

  a. pipe wrench   **b.** emergency call   **c.** fixing pipes   **d.** drains

USE YOUR OWN WORDS

>>>> *Look at the picture. What words come into your mind other than the ten vocabulary words used in this lesson? Write them on the lines below. To help you get started, here are two good words:*

1. _____dramatic_____
2. _____exciting_____
3. _____
4. _____
5. _____
6. _____
7. _____
8. _____
9. _____
10. _____

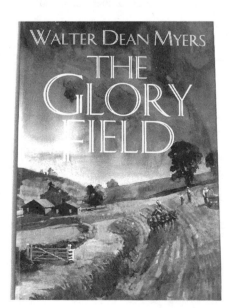

UNSCRAMBLE THE LETTERS

>>>> *Each group of letters represents one of the vocabulary words for this lesson. Can you unscramble them? Write your answers in the blanks on the right.*

Scrambled Letters   **Vocabulary Words**

1. slenul   _____

2. finallymor   _____

3. lovens   _____

4. myries   _____

5. reem   _____

6. driftel   _____

7. bilioncaput   _____

8. atled   _____

9. sribarer   _____

10. fededin   _____

COMPLETE THE STORY

>>>> Here are the ten vocabulary words for this lesson:

novels	misery	barriers	flirted	mere
sullen	define	publication	dealt	informally

>>>> *There are six blank spaces in the story below. Four vocabulary words have already been used in the story. They are underlined. Use the other six words to fill in the blanks.*

When you talk <u>informally</u> with friends about books, do you discuss _____ or nonfiction? People <u>define</u> a book of nonfiction as a _____ that is true and based on facts. When you are assigned a book report at school, do you feel joy or _____? Some students look <u>sullen</u> when they are asked to read a book they do not find interesting. They may spend a _____ hour or less skimming the book.

However, young people want to read Walter Dean Myers's books because they are about teenagers who have _____ with danger and even death. Some of his characters have _____ with major hardships and <u>barriers</u>. They have found ways to survive and even to succeed.

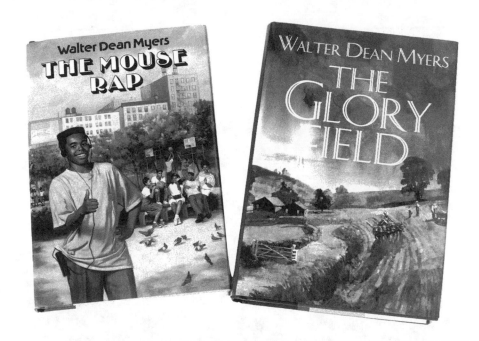

Learn More About Writers

>>>> *On a separate sheet of paper or in your notebook or journal, complete one or more of the activities below.*

Learning Across the Curriculum

Write a short story or create some kind of visual art based on your own childhood memories and experiences. If you wish, share your work with a partner or a small group.

Broadening Your Understanding

Read a novel by an author you enjoy. Then, based on the novel, write an essay on how you think the author's childhood experiences may have influenced the story.

Extending Your Reading

Choose one of the following books by Walter Dean Myers. After reading the book, explain how the story might change if it had a different setting. For example, suppose the main characters lived or grew up in the suburbs or a farming community instead of the city. How might their lives and their problems change? What parts of the story might stay the same?

Glory Field
The Mouse Rap
Fast Sam, Cool Clyde, and Stuff
Hoops
It Ain't All for Nothin'
Somewhere in the Darkness
Motown and Didi
Scorpions
The Young Landlords

5 THE GOLDEN CITY

Although this city has been invaded and destroyed or partly destroyed more than 40 times, it has always been rebuilt. Ruin has been piled upon ruin until today the streets are 35 feet higher than they were 2,000 years ago.

For ages, Jerusalem has been a $\boxed{sacred}$ city. It is the center of three $\boxed{religious}$ groups—the Jews, the Christians, and the Muslims. For this reason, Jerusalem has been called the land of the star, the cross, and the $\boxed{crescent.}$ Each of these three $\boxed{symbols}$ stands for one of the three religions. The star is the Star of David for the Jewish religion. The cross is the cross of Jesus for the Christian religion. The crescent is the symbol for the Muslim religion.

Jerusalem is divided into two parts—the old city and the new city. The new Jerusalem $\boxed{resembles}$ a modern city. It has tall buildings and crowded streets. The old Jerusalem has hardly changed during its long history. Many of the towers and religious $\boxed{shrines}$ that were built long ago still stand.

The clothes worn by the people of Jerusalem also $\boxed{reflect}$ the old and the new. Some people wear the $\boxed{apparel}$ of their $\boxed{forebears.}$ Others dress in the latest styles.

New and old, Jews, Christians, and Muslims, all $\boxed{mingle}$ in the narrow streets of Jerusalem. Both the beauty and importance of this great city have made it "Jerusalem the Golden."

UNDERSTANDING THE STORY

 Circle the letter next to each correct statement.

1. The main idea of this story is that
   a. Jerusalem is a city torn by wars and in need of rebuilding.
   b. Jerusalem is a mix of the old and the new, and its people are a mix of different faiths.
   c. symbols are necessary to tell religions apart.

2. From this story, you can conclude that
   a. millions of people have strong emotional ties to Jerusalem.
   b. the days of the destruction of Jerusalem are over.
   c. people in the new part of the city are more religious than people in the old part.

MAKE AN ALPHABETICAL LIST

>>>> *Here are the ten vocabulary words in this lesson. Write them in alphabetical order in the spaces below.*

sacred	religious	crescent	symbols	shrines
resembles	apparel	reflect	forebears	mingle

1. _____
2. _____
3. _____
4. _____
5. _____

6. _____
7. _____
8. _____
9. _____
10. _____

WHAT DO THE WORDS MEAN?

>>>> *Following are some meanings, or definitions, for the ten vocabulary words in this lesson. Write the words next to their definitions.*

1. _____ holy; worthy of reverence

2. _____ the shape of the moon in the first or last quarter; the symbol of the Muslim religion

3. _____ things that stand for or represent something else; signs

4. _____ having to do with a belief in God; devout

5. _____ looks like; is similar in appearance

6. _____ sacred places; places where holy things are kept

7. _____ to give back an image of

8. _____ clothing; dress

9. _____ family members who lived a long time ago

10. _____ to mix; to get along together

COMPLETE THE SENTENCES

>>>> *Use the vocabulary words in this lesson to complete the following sentences. Use each word only once.*

shrines	apparel	resembles	symbols	religious
crescent	sacred	mingle	reflect	forebears

1. Jerusalem is a city that has _____ meaning for people of different faiths.

2. In buildings that were used by their _____, the people worship daily.

3. Throughout Jerusalem, one sees three famous _____—the star, the cross, and the crescent.

4. The _____ is a symbol that has special meaning for the Muslim people.

5. The Western Wall is a structure that is _____ to the Jewish people.

6. The religious _____ in the old city are visited by people from far and near.

7. People of different races, religions, and styles _____ in the streets.

8. To judge by the _____ worn by some of the people in the old city, there has been little change over the centuries.

9. However, the new part of Jerusalem _____ most other modern cities.

10. The tall buildings in new Jerusalem _____ the modern tastes of people in that section of the city.

USE YOUR OWN WORDS

>>>> *Look at the picture. What words come into your mind other than the ten vocabulary words used in this lesson? Write them on the lines below. To help you get started, here are two good words:*

1. _____dome_____
2. _____trucks_____
3. _____
4. _____
5. _____
6. _____
7. _____
8. _____
9. _____
10. _____

FIND THE ANALOGIES

>>>> In an **analogy,** similar relationships occur between words that are different. For example, *pig* is to *hog* as *car* is to *automobile*. The relationship is that the words mean the same. Here's another analogy: *noisy* is to *quiet* as *short* is to *tall*. In this relationship, the words have opposite meanings.

>>>> *See if you can complete the following analogies. Circle the correct word or words.*

1. **Crescent** is to **Muslim** as **star** is to

a. sky   **b.** Christian   **c.** astronaut   **d.** Jewish

2. **Holy** is to **sacred** as **devout** is to

a. man   **b.** child   **c.** religious   **d.** shrines

3. **Knowledge** is to **ignorance** as **mingle** is to

a. mix   **b.** together   **c.** separate   **d.** truth

4. **Apparel** is to **body** as **shoe** is to

a. shine   **b.** laces   **c.** leather   **d.** foot

5. **Sacred** is to **unholy** as **up** is to

a. planes   **b.** down   **c.** sad   **d.** planets

COMPLETE THE STORY

>>>> Here are the ten vocabulary words for this lesson:

sacred	religious	crescent	symbols	shrines
resembles	apparel	reflect	forebears	mingle

>>>> *There are six blank spaces in the story below. Four vocabulary words have already been used in the story. They are underlined. Use the other six words to fill in the blanks.*

Within old Jerusalem are three famous _____ that have been sacred to Christians, Jews, and Muslims for many centuries. While new Jerusalem _____ any other modern city, the old city of the star, the cross, and the crescent has changed very little. These symbols _____ the many influences upon this city.

Its narrow, winding streets are filled with people wearing the same kind of _____ that their forebears wore. Many very _____ people of different traditions meet and _____ with one another. If you visit this city, you can't help being touched by its history and culture.

30

Learn More About the Middle East

>>>> *On a separate sheet of paper or in your notebook or journal, complete one or more of the activities below.*

Appreciating Diversity

Learn more about a large city in another country. Research what it is like and write a report. You may want to illustrate your report with your own art.

Broadening Your Understanding

Find out more about what tourists go to see in Jerusalem. Then read more about these sights. Now imagine you have just been to visit Jerusalem. Write a postcard to a friend at home explaining the most interesting sight. Draw a picture of what you saw for the front of the postcard.

Learning Across the Curriculum

Read the newspaper and find an article about the Middle East. Find more about the issue the newspaper article is addressing. Then write a summary of what you learn.

6 A BASEBALL LEGEND

As a small boy in Puerto Rico, Roberto Clemente dreamed of playing baseball. He did not plan to become one of the most admired players in history. He just wanted to play the game he loved so much.

One day, a baseball scout from the United States recognized young Clemente's great athletic skill. He took Clemente to the United States, where the young man ended up playing for the Pittsburgh Pirates. At first, Clemente experienced *prejudice* because of his skin color and his *cultural* background. He also had problems speaking English.

Clemente was proud to be Puerto Rican, however, and made no *apology* for his background. *Despite* insults from a few people, Clemente did not carry a *grudge.* He *persisted* in doing his best and helped the Pirates win the 1960 and 1971 World Series. He also earned four National League batting titles and was twice voted Most Valuable Player. Clemente's amazing skill brought him fame and admiration.

Still, this *recognition* did not make Clemente forget his *obligations* to his fans. For example, after winning the 1960 World Series, Clemente went back to the field to hug his delighted fans. Someone once gave Clemente $6,000. He donated it to Children's Hospital in Pittsburgh.

Clemente's concern for the *welfare* of others led to his death. In 1972, a strong earthquake left thousands in Nicaragua dead, hurt, or homeless. Clemente was in *anguish* over the situation. He decided to fly food and supplies to Nicaragua. Shortly after taking off, the plane plunged into the sea with Clemente aboard. No one survived.

Clemente is remembered not just as a baseball superstar, but also as a great and caring man.

UNDERSTANDING THE STORY

>>>> *Circle the letter next to each correct statement.*

1. The statement that best expresses the main idea of this selection is that
   a. Clemente became famous and successful, and he reached out and helped others.
   b. Clemente helped others in order to draw attention to himself.
   c. Clemente was wrong to care about others.

2. From this story, you can conclude that
   a. Clemente's family encouraged him to be a baseball superstar.
   b. Clemente was ashamed of his background.
   c. Clemente did not mind when fans asked for his autograph.

MAKE AN ALPHABETICAL LIST

>>>> *Here are the ten vocabulary words in this lesson. Write them in alphabetical order in the spaces below.*

| recognition | welfare | despite | persisted | grudge |
| apology | anguish | obligations | prejudice | cultural |

1. _____   6. _____

2. _____   7. _____

3. _____   8. _____

4. _____   9. _____

5. _____   10. _____

WHAT DO THE WORDS MEAN?

>>>> *Following are some meanings, or definitions, for the ten vocabulary words in this lesson. Write the words next to their definitions.*

1. _____ an expression of regret for wrongdoing

2. _____ happiness; well being

3. _____ resentment; ill feelings

4. _____ pain; sorrow

5. _____ continued in spite of obstacles

6. _____ special notice or attention

7. _____ dislike of people who are different

8. _____ duties; responsibilities

9. _____ not prevented by; in spite of

10. _____ relating to the beliefs and behaviors of a social, ethnic, or religious group

FIND THE ANALOGIES

>>>> An **analogy** is a relationship between words. Here's one kind of analogy: *rain* is to *flood* as *sun* is to *burn*. In this relationship, the first word in each pair is the cause and the second word in each pair is the effect.

>>>> *See if you can complete the following analogies. Circle the correct word or words.*

1. **Grudge** is to **resentment** as **failure** is to

   a. ambition   **b.** disappointment   **c.** contentment   **d.** attempt

2. **Winning** is to **recognition** as **practicing** is to

   a. improvement   **b.** awards   **c.** failure   **d.** giving up

3. A **hard decision** is to **anguish** as a **funny movie** is to

   a. sorrow   **b.** nail biting   **c.** laughter   **d.** crying

4. A **job** is to **obligations** as **a day off** is to

   a. relaxation   **b.** responsibility   **c.** tension   **d.** resentment

5. **Making a mistake** is to **apology** as **receiving help** is to

   a. forgiven   **b.** anguish   **c.** congratulations   **d.** thanks

USE YOUR OWN WORDS

>>>> *Look at the picture. What words come into your mind other than the ten vocabulary words used in this lesson? Write them on the blank lines below. To help you get started, here are two good words:*

1. _____ admiration _____
2. _____ caring _____
3. _____
4. _____
5. _____
6. _____
7. _____
8. _____
9. _____
10. _____

DESCRIBE THE NOUNS

>>>> *Two of the vocabulary words, obligations and anguish, are nouns. List as many words as you can that describe or tell something about the words obligations and anguish. You can work on this with your classmates.*

obligations

1. _____
2. _____
3. _____
4. _____
5. _____
6. _____
7. _____
8. _____
9. _____
10. _____

anguish

1. _____
2. _____
3. _____
4. _____
5. _____
6. _____
7. _____
8. _____
9. _____
10. _____

COMPLETE THE STORY

>>>> Here are the ten vocabulary words for this lesson:

apology	despite	welfare	grudge	obligations
anguish	cultural	recognition	prejudice	persisted

>>>> *There are six blank spaces in the story below. Four vocabulary words have already been used in the story. They are underlined. Use the other six words to fill in the blanks.*

Many athletes have faced _____ because of their <u>cultural</u> backgrounds. Some carry a _____ throughout their careers. Others put aside their <u>anguish</u> and win respect and _____ with their athletic skill. <u>Despite</u> the fact that some fans insulted him, Roberto Clemente continued to care about the _____ of all people.

Even when he was famous and popular, he felt he had <u>obligations</u> to others. He _____ in reaching out to people in trouble. Clemente never needed to make an _____ for the way he led his short life.

36

Learn More About Sports

>>>> *On a separate sheet of paper or in your notebook or journal, complete one or more of the activities below.*

Building Language

Explain at least two meanings for each of these words: *fan, sport, field.* Use each word in two sentences to show the different meanings.

Learning Across the Curriculum

Write a short report about another sports star who is admired for his or her contributions outside of sports. Choose two names from the list below. Share what you learn with the class.

Wilma Rudolph
Bill Bradley
Arthur Ashe
Jackie Joyner-Kersee
José Torres

Broadening Your Understanding

In a short report, compare Roberto Clemente with one of today's baseball players. How are their attitudes toward playing baseball and toward their fans similar? How are they different? How would you explain the differences in their attitudes?

Country songs often tell of somebody's troubles. When Randy Travis sings, the words have real meaning for him. He was born near Marshville, North Carolina. His family worked hard, but his father urged his boys to take up the guitar. Travis practiced and gave his first performance at the age of eight.

But Travis got into trouble. He dropped out of school in the ninth grade. He argued with his parents and was in constant conflict. He was even arrested five times for drug- and drink-related incidents. Today he shrugs and says, "Now, I wouldn't have the ignorance or the nerve to do those things."

His life was altered when he met Lib Hatcher. She owned a restaurant. She hired Randy to cook, clean up, and sing for the customers. During this period, she helped him straighten out his life. One night, a talent scout from Warner Records heard Travis sing. She immediately gave him a contract. His first record was released twice. The first time, it didn't even make the top 50. The second time, it soared to number 1 on the charts.

Today Travis travels much of the year. Hatcher is his business manager and his best friend. Travis likes to perform for his fans. "They feel like family when they like you," he admits, with a shy smile. Travis still has one goal. He has received numerous awards for his singing, but he wants to be recognized as a songwriter, too. Today most of his troubles seem to be the ones expressed in his songs.

UNDERSTANDING THE STORY

 Circle the letter next to each correct statement.

1. The main idea of this story is that
   a. Randy Travis was a wild teenager.
   b. country music is popular.
   c. country music changed Travis's life.

2. Randy Travis knows what country songs are about because
   a. his own life has been easy and unhurried.
   b. his early life was full of trouble and pain.
   c. he was born in North Carolina.

MAKE AN ALPHABETICAL LIST

>>>> *Here are the ten vocabulary words in this lesson. Write them in alphabetical order in the spaces below.*

urged	contract	arrested	manager	ignorance
recognized	altered	soared	talent	shy

1. _____

2. _____

3. _____

4. _____

5. _____

6. _____

7. _____

8. _____

9. _____

10. _____

WHAT DO THE WORDS MEAN?

>>>> *Following are some meanings, or definitions, for the ten vocabulary words in this lesson. Write the words next to their definitions.*

1. _____ a lack of knowledge

2. _____ a legal paper promising a job

3. _____ a natural gift for doing something

4. _____ identified

5. _____ a performer's business arranger

6. _____ held by the police

7. _____ advised strongly

8. _____ changed

9. _____ modest; uncertain

10. _____ rose upward quickly

COMPLETE THE SENTENCES

>>>> *Use the vocabulary words in this lesson to complete the following sentences. Use each word only once.*

urged	contract	arrested	manager	ignorance
recognized	altered	soared	talent	shy

1. Time has _____ Randy Travis's feelings about life.

2. He thinks his _____ got him into trouble.

3. He was _____, but he never went to prison.

4. He has a natural _____ for singing country songs.

5. The woman who helped Travis change his life is now his _____.

6. An agent from a large record company offered Travis a _____.

7. Travis is _____ when he talks about his success as a singer.

8. He would really like to be _____ as a songwriter.

9. He is still glad that his father _____ him to practice the guitar.

10. Travis's popularity has _____.

USE YOUR OWN WORDS

>>>> *Look at the picture. What words come into your mind other than the ten vocabulary words used in this lesson? Write them on the blank lines below. To help you get started, here are two good words:*

1. _____ cheerful _____
2. _____ performer _____
3. _____
4. _____
5. _____
6. _____
7. _____
8. _____
9. _____
10. _____

FIND THE ANTONYMS

>>>> **Antonyms** are words that are opposite in meaning. For example, *good* and *bad* and *fast* and *slow* are antonyms. Here are antonyms for six of the vocabulary words.

>>>> *See if you can find the vocabulary words and write them in the blanks on the left.*

Vocabulary Word	Antonym
1. _____	unchanged
2. _____	boastful
3. _____	wisdom
4. _____	dropped
5. _____	ignored
6. _____	discouraged

COMPLETE THE STORY

>>>> Here are the ten vocabulary words for this lesson:

urged	contract	arrested	manager	ignorance
recognized	altered	soared	talent	shy

>>>> *There are six blank spaces in the story below. Four vocabulary words have already been used in the story. They are underlined. Use the other six words to fill in the blanks.*

Randy Travis is _____ in interviews. He seems surprised that people recognize him. He talks about being _____ and about other teenage troubles. He says that his <u>ignorance</u> got him into trouble. Lib Hatcher _____ his life. She gave him his first job. She _____ him to sing for her customers. She <u>recognized</u> his _____ as a country singer. That job led to a recording <u>contract</u>. Hatcher is his business _____ now. His popularity <u>soared</u>, but Travis still thinks of himself as just a country boy.

Learn More About Country Music

>>>> *On a separate sheet of paper or in your notebook or journal, complete one or more of the activities below.*

Building Language

Listen to a recording of a country song. Ask a friend to help you if you cannot understand some of the language. Then write what you think the song is saying and what the singer wants to communicate.

Learning Across the Curriculum

Country music has a long and fascinating history in this country. Read an account of how country-and-western music began. Share your information with the class.

Broadening Your Understanding

Locate a song book with country-and-western songs. Read the lyrics. Then write a paragraph explaining what these songs are about. Are there some themes that seem to come up again and again in country-and-western music? Explain what these themes are in an oral presentation.

8 BLUE WHALES

Whales are among the largest, most powerful animals that have ever lived. Some dinosaurs were small by comparison. The blue whale, the largest of all whales, can grow to more than 100 feet and weigh 150 tons. Despite its **tremendous** size, the blue whale is **vulnerable** to hunters.

The blue whale was first hunted by the North American Inuit (Eskimos). In fact, whales were **essential** to the Inuit's existence. For centuries, they hunted these huge **mammals.** They used the whales for food, but nothing was wasted. The whale oil became fuel. The sinews became ropes. The bones were used as tools.

Today the blue whale is still hunted by some nations. Its oil is used to make soap and other products. But there are alternative sources for all of these products. People no longer need to hunt blue whales to survive.

Modern-day whalers use advanced **apparatus.** Radar and helicopters are used to find the whales. Deadly **harpoon** guns are used to kill them. The **carnage** resulting from the use of this equipment has brought the blue whale close to extinction.

Conservationists have warned us. If we allow whalers to continue in this way, **extermination** of the blue whale is almost certain.

It would be a shameful loss if the number of blue whales **dwindled** to nothing. The blue whale can never be replaced.

UNDERSTANDING THE STORY

 Circle the letter next to each correct statement.

1. The main purpose of this story is to
   a. describe the features of the blue whale.
   b. warn readers that the blue whale is in danger of becoming extinct.
   c. describe the many ways the Inuit used whales in their daily life.

2. From this story, you can conclude that
   a. it is only a rumor that the blue whale is in danger of becoming extinct.
   b. the blue whale has learned to avoid the deadly harpoon gun.
   c. the public must do something soon if the blue whale is to survive.

MAKE AN ALPHABETICAL LIST

>>>> *Here are the ten vocabulary words in this lesson. Write them in alphabetical order in the spaces below.*

tremendous	extermination	mammals	essential	harpoon
apparatus	conservationists	carnage	dwindled	vulnerable

1. _____

2. _____

3. _____

4. _____

5. _____

6. _____

7. _____

8. _____

9. _____

10. _____

WHAT DO THE WORDS MEAN?

>>>> *Following are some meanings, or definitions, for the ten vocabulary words in this lesson. Write the words next to their definitions.*

1. _____ huge; enormous

2. _____ the killing of a great number of people or animals

3. _____ animals that feed milk to their young; people belong to this group

4. _____ absolutely necessary

5. _____ materials, tools, special instruments, or machinery needed to carry out a purpose

6. _____ persons who wish to save forms of animal and plant life that are in danger of being destroyed forever

7. _____ a long spear with a rope tied to it used in killing a whale

8. _____ defenseless against; open to attack or injury

9. _____ the act of destroying completely; putting an end to

10. _____ reduced in number

46

COMPLETE THE SENTENCES

>>>> *Use the vocabulary words in this lesson to complete the following sentences. Use each word only once.*

apparatus	dwindled	harpoon	mammals	extermination
essential	tremendous	carnage	vulnerable	conservationists

1. Not enough is being done to prevent the _____ of the blue whale.

2. Whales have become more _____ to extermination.

3. In some areas, the number of blue whales has _____ to only a few dozen.

4. Animals that feed milk to their young are called _____.

5. The _____ used to hunt whales today is very advanced.

6. The whale was _____ to the Inuit's existence.

7. The _____ of the whale that still goes on today is harder to excuse.

8. The _____ that used to be thrown by a person is now shot out of a gun.

9. _____ are calling on people to outlaw the killing of whales.

10. Blue whales are _____ in size.

USE YOUR OWN WORDS

>>>> *Look at the picture. What words come into your mind other than the ten vocabulary words used in this lesson? Write them on the lines below. To help you get started, here are two good words:*

1. _____ water _____

2. _____ splash _____

3. _____

4. _____

5. _____

6. _____

7. _____

8. _____

9. _____

10. _____

MAKE POSSESSIVE WORDS

>>>> The possessive of a word shows that something belongs to it. For example, Bill has a boat; it is *Bill's* boat. To make a possessive of a word that doesn't end in *s*, add an apostrophe and an *s* to the word, such as *baker's* bread or *father's* car. To make a possessive of a word that does end in *s*, add an apostrophe (*s'*), such as *friends'* bicycles or *ladies'* hats.

>>>> *Here are ten words from the story. In the space next to the word, write the correct possessive of the word.*

1. ships _____

2. fleet _____

3. ocean _____

4. bodies _____

5. whale _____

6. Inuit _____

7. whalers _____

8. conservationists _____

9. hunter _____

10. boats _____

COMPLETE THE STORY

>>>> Here are the ten vocabulary words for this lesson:

extermination	conservationists	tremendous	essential	harpoon
carnage	apparatus	vulnerable	dwindled	mammals

>>>> *There are six blank spaces in the story below. Four vocabulary words have already been used in the story. They are underlined. Use the other six words to fill in the blanks.*

The blue whale is the largest animal that has ever lived. However, _____ are worried. If whalers keep killing these huge _____, the blue whale is doomed to <u>extermination</u>.

Sixty years ago, there were many blue whales in the world. Today the number has _____ to very few. The deadly _____ gun is the major reason for this. Whalers have more modern <u>apparatus</u> to help them find and kill the whales. The common use of helicopters and radar makes the whales very _____.

It is <u>essential</u> that the killing of this _____ animal be stopped. If the <u>carnage</u> continues, the blue whale will completely disappear from the oceans of the world.

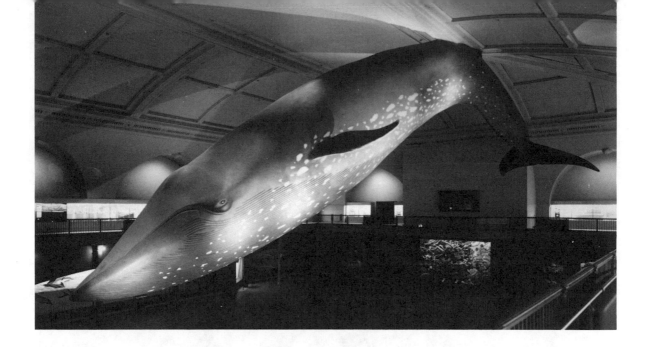

Learn More About Endangered Species

>>>> *On a separate sheet of paper or in your notebook or journal, complete one or more of the activities below.*

Learning Across the Curriculum

Why do animals become endangered or extinct? What can we do to stop these animals from dying out? Do some research to find out. Then write a children's book that explains the problems of endangered species and what we can do to solve this problem.

Broadening Your Understanding

Choose an animal that has become extinct. Find out more about it and write a report on the animal. Explain what happened to it and why. If you can, include a drawing or an illustration of the animal you wrote about.

Extending Your Reading

What is being done today to protect whales from extinction? Read one of these books to find out. Then write a paragraph that explains what you discovered, or tell one specific way in which whales are being protected.

Saving the Whales, by Michael Bright
Whales and Dolphins, by Steve Parker
Humpback Whale, by Michael Bright
All About Whales, by Deborah Kovacs

9 THE MARY ROSE

In 1545, the pride of the English navy, the *Mary Rose,* set sail. Its mission was to fight the French. Among those watching from shore was the monarch, Henry VIII. The crowd was cheering when the ship left Portsmouth harbor. The outcome, however, was disaster. The 700-ton *Mary Rose* was laden with cannons. Before it got far from shore, it sank. Historians say the cannons were not properly bolted. The loose cannons rolled across the deck and through the ship's side. In rushed the sea. More than 650 sailors lost their lives. It was a grave blow to the English people and their king.

More than 420 years later, the *Mary Rose* was found. A diver, Alexander McKee, discovered the wreck a mile off the coast of England. For four years, McKee and his friends worked to clear away mud and silt. They finally brought up a cannon. This find stirred the public's interest. Money was contributed to raise the *Mary Rose*. Prince Charles became president of the Mary Rose Trust, which raised funds.

The salvage continued during the summer of 1982. With the use of a special lifting crane and cradle, the *Mary Rose* was brought to the surface. Most of its oak frame was still in fair condition. To prevent further decay, the hull was wrapped in plastic sheeting. The rescue marked the end of a 17-year-long project. The cost had been $7 million, but it was worth every penny. The restored *Mary Rose* is now on permanent display in England.

UNDERSTANDING THE STORY

>>>> *Circle the letter next to each correct statement.*

1. The main idea of this story is that
   a. it took $7 million to raise the *Mary Rose*.
   b. the *Mary Rose* sank without firing a shot at the enemy.
   c. after more than 400 years at the bottom of the sea, the *Mary Rose* was salvaged.

2. From this story, you can conclude that
   a. Prince Charles will continue to search for lost wrecks.
   b. from now on, cannons aboard ships will be more securely bolted.
   c. the British people felt great pride in the rescue of the *Mary Rose*.

MAKE AN ALPHABETICAL LIST

>>>> *Here are the ten vocabulary words in this lesson. Write them in alphabetical order in the spaces below.*

plastic	monarch	mission	bolted	laden
cradle	restored	salvage	project	grave

1. _____

2. _____

3. _____

4. _____

5. _____

6. _____

7. _____

8. _____

9. _____

10. _____

WHAT DO THE WORDS MEAN?

>>>> *Following are some meanings, or definitions, for the ten vocabulary words in this lesson. Write the words next to their definitions.*

1. _____ an undertaking; often a big, complicated job

2. _____ a king or queen; an absolute ruler

3. _____ a framework upon which a ship rests, usually during repair

4. _____ fastened; held with metal fittings

5. _____ a special task

6. _____ a synthetic or processed material

7. _____ serious; critical

8. _____ the act of saving a ship or its cargo from the sea

9. _____ loaded; heavily burdened

10. _____ brought back to its original state; reconstructed

COMPLETE THE SENTENCES

>>>> *Use the vocabulary words in this lesson to complete the following sentences. Use each word only once.*

bolted	cradle	plastic	grave	monarch
project	laden	salvage	restored	mission

1. We knew that the _____ was successful when we saw the ship's mast.

2. The cargo was securely _____ to the deck so that it wouldn't roll.

3. It may take three years for the ship to be _____ to its former glory.

4. The king said, "Your _____ is to seek out and destroy the enemy."

5. Little did the _____ know that his favorite ship would soon sink.

6. The problem was that the ship was _____ with heavy cannons.

7. The _____ involved hundreds of people and millions of dollars.

8. A steel _____ was designed to support the waterlogged ship.

9. The loss of 650 sailors was a _____ blow to the English people.

10. _____ was wrapped around the decaying hull to keep it from the air.

USE YOUR OWN WORDS

>>>> *Look at the picture. What words come into your mind other than the ten vocabulary words used in this lesson? Write them on the lines below. To help you get started, here are two good words:*

1. _____mechanical_____

2. _____calm_____

3. _____

4. _____

5. _____

6. _____

7. _____

8. _____

9. _____

10. _____

MATCH TERMS WITH THEIR MEANINGS

>>>> *Here are some words selected from the world of sailing and ships. See if you can match the terms with their meanings. You may need the help of a dictionary.*

1. **hull** _____   **a.** the left-hand side of a ship

2. **port** _____   **b.** a long pole holding sails

3. **starboard** _____   **c.** the vertical blade at the rear of a ship, used to change course

4. **stern** _____   **d.** the body of a ship

5. **mast** _____   **e.** having to do with ships and the sea

6. **rudder** _____   **f.** the back or rear of a ship

7. **nautical** _____   **g.** the right-hand side of a ship

COMPLETE THE STORY

>>>> Here are the ten vocabulary words for this lesson:

monarch	salvage	restored	bolted	project
grave	laden	plastic	cradle	mission

>>>> *There are six blank spaces in the story below. Four vocabulary words have already been used in the story. They are underlined. Use the other six words to fill in the blanks.*

The _____ Henry VIII was in a good mood. His fleet was on its way to fight the French. It was a <u>mission</u> that he strongly supported. He was particularly proud of the flagship, the *Mary Rose*. It was _____ with 91 cannons. What firepower!

Then suddenly everything went wrong. The heavy cannons were not _____ well enough. They rolled across the deck and crashed through the ship's side. The ship went down. Many sailors were drowned. What a <u>grave</u> loss to the nation! But 437 years later, the *Mary Rose* was <u>restored</u> to life. A _____ brought the *Mary Rose* to the surface. It was a huge <u>project</u> but it succeeded. A special steel _____ was built to raise the hull. As soon as the hull reappeared, it was wrapped in _____ to prevent further decay. The *Mary Rose* was home again.

Learn More About Sailing Ships

>>>> *On a separate sheet of paper or in your notebook or journal, complete one or more of the activities below.*

Learning Across the Curriculum

How did early sailors know where they were going? Ancient people who sailed the seas invented many devices to help them navigate. Among these are the kamal, the astrolabe, the sextant, the magnetic compass, and the sundial. Find out more about one of these devices. In a paragraph or two, explain how it worked. If you can, make a model of it to show to the class.

Broadening Your Understanding

Research life on sailing ships in the 1500s. You can also look for information about life on the *Mary Rose*. Both *Time* and *Newsweek* magazines had stories about the salvage that raised the ship. Write an account of what you think life was like for the sailors on board the ship. You may want to write your account as if you were a sailor on the ship keeping a log, or diary.

Extending Your Reading

Create a timeline of the history of sailing ships. The books below will aid your research:

Ships, Sailors and the Sea, by Richard Humble
Ships Come Aboard, by Seigfried Aust

10 BIGFOOT

It is the first night of your camping trip. You are sitting with your friends around the campfire. You think back to the day's activities. You were *fascinated* by all the *spectacular* sights.

Your guide interrupts your thoughts and begins to tell a chilling story. "A gigantic, wild, hairy beast that looks and walks like a man roams all over this country," the guide says. "Many *reputable* people claim to have seen it. One rancher says he has taken motion pictures of it. No one knows where this creature comes from or where it goes. No one has been able to *identify* it. The creature is called Bigfoot."

You are *skeptical* about the whole story. "It makes good *fiction,*" you think to yourself, "but it couldn't be true." Later, as you begin to fall asleep, you decide the story is nothing but a *hoax.*

Soon you are jolted awake by the shouts of one of your companions. "There it is!" You quickly look in the direction he is pointing. A huge animal, covered with dark hair, is coming toward you. But the shouts scare the *primitive* beast away.

The next morning, you think there must be an *explanation* for what happened. Was it a dream? But soon you find the *proof* you need—trampled grass and a few broken branches. No, it wasn't a dream. As you get ready for breakfast, you wonder, "Have I seen Bigfoot?"

UNDERSTANDING THE STORY

 Circle the letter next to each correct statement.

1. The main purpose of this story is to
   a. cause the reader to wonder if Bigfoot really exists.
   b. show how exciting camping can be with the right people along.
   c. warn the reader about the dangers of camping.

2. From this story, you can conclude that
   a. the story of Bigfoot has finally been proved untrue.
   b. there will probably always be reports of seeing Bigfoot.
   c. Bigfoot is really a huge bear.

MAKE AN ALPHABETICAL LIST

>>>> *Here are the ten vocabulary words in this lesson. Write them in alphabetical order in the spaces below.*

fascinated	spectacular	reputable	identify	skeptical
fiction	hoax	primitive	explanation	proof

1. _____
2. _____
3. _____
4. _____
5. _____

6. _____
7. _____
8. _____
9. _____
10. _____

WHAT DO THE WORDS MEAN?

>>>> *Following are some meanings, or definitions, for the ten vocabulary words in this lesson. Write the words next to their definitions.*

1. _____ having doubts; not willing to believe

2. _____ a trick

3. _____ to recognize as being a particular person or thing

4. _____ amazed; very interested by

5. _____ a story that is not true

6. _____ facts; evidence

7. _____ honorable; well thought of

8. _____ eye-catching; very unusual

9. _____ living long ago; from earliest times

10. _____ a statement that clears up a difficulty or a mistake

COMPLETE THE SENTENCES

>>>> *Use the vocabulary words in this lesson to complete the following sentences. Use each word only once.*

fiction	spectacular	skeptical	hoax	proof
primitive	explanation	identify	fascinated	reputable

1. People have always been _____ by stories of strange creatures.

2. The story of Bigfoot seems closer to science _____ than to fact.

3. A professor called the Bigfoot story a _____.

4. She said that she needed more _____, such as a clear photograph.

5. Her _____ of broken branches wasn't convincing.

6. She wasn't able to _____ a piece of hair found clinging to a branch.

7. Those who claim to have seen Bigfoot seem _____.

8. The camper saw many _____ sights that day.

9. The police were _____ when two people in different places reported seeing Bigfoot at the same time.

10. The monster is called _____ because it seems like something prehistoric.

USE YOUR OWN WORDS

>>>> *Look at the picture. What words come into your mind other than the ten vocabulary words used in this lesson? Write them on the lines below. To help you get started, here are two good words:*

1. _____ leaves _____
2. _____ hairy _____
3. _____
4. _____
5. _____
6. _____
7. _____
8. _____
9. _____
10. _____

>>>> *Look at the vocabulary word below. See how many words you can form by using the letters of this word. Make up at least ten words. One has already been done for you. Write your words in the spaces below.*

reputable

1. _____ plate _____   7. _____

2. _____   8. _____

3. _____   9. _____

4. _____   10. _____

5. _____   11. _____

6. _____   12. _____

COMPLETE THE STORY

>>>> Here are the ten vocabulary words for this lesson:

fascinated	spectacular	reputable	identify	skeptical
fiction	hoax	primitive	explanation	proof

>>>> *There are six blank spaces in the story below. Four vocabulary words have already been used in the story. They are underlined. Use the other six words to fill in the blanks.*

Many _____ people claim they have seen Bigfoot, although scientists are _____ about this beast. Some experts think Bigfoot is nothing but a <u>hoax</u>.

One thing is sure: Huge footprints have been found in the areas where the <u>primitive</u> creature roams. No one has been able to _____ these <u>spectacular</u> tracks. No one has ever seen anything like them before.

Bigfoot has _____ many people. Some believe Bigfoot is real; others say it is just _____. Scientists hope to gather more _____ about Bigfoot. Then, hopefully, we will have an <u>explanation</u> about this strange and mysterious creature.

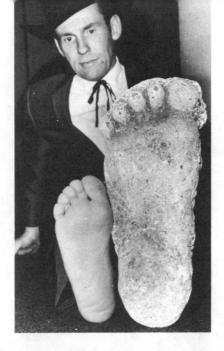

Learn More About Monsters

>>>> *On a separate sheet of paper or in your notebook or journal, complete one or more of the activities below.*

Appreciating Diversity

Monsters, real or imaginary, are a part of many cultures. Think back to your childhood and remember the frightening stories you were told. Write one down. How is it like the Bigfoot story? How is it different?

Broadening Your Understanding

Find out more about Bigfoot. Research this creature. Write a report arguing that Bigfoot either does or does not exist. Find a friend with the opposite opinion. Have a debate with your friend about the existence of Bigfoot. Have the class vote on whose argument is more convincing.

Extending Your Reading

Read one of the books below about movie monsters. Then watch a movie about a monster, such as Godzilla. Finally, write a paragraph that explains why people think the monster is frightening.

Horror in the Movies, by Daniel Cohen
Frankenstein, by John Turvey
Movie Monsters, by Thomas G. Aylesworth
Dracula, by Ian Thorne

11 THE LEARNING MAN

Gordon Parks is a *notable* photographer. People love his pictures because he uses his camera the way a painter uses a brush.

Parks learned about photography at a Chicago art center. He became *enthusiastic* about the idea of *pursuing* a career in this field. To earn money, he worked as a waiter, a lumberjack, and a piano player. He led a band. He even played baseball.

Everyone saw that Parks's talent was *considerable.* He won a *scholarship.* Now he had an opportunity to study without worrying about money.

In time, Parks became a top magazine photographer. He traveled all over the world, creating wonderful stories in pictures. Among his *assignments* were stories on *segregation* and crime. He also wrote a report on the life of a gang leader in New York's Harlem. He also made a *documentary* of his story on the *plight* of a poor boy in Brazil.

The accomplishments of this man are many. Parks is known for his original music. He is a master of the documentary. His novel *The Learning Tree* is based on his life. It was made into a movie that Parks produced, directed, and photographed. In 1989, *The Learning Tree* was declared one of the national treasures of American film.

Gordon Parks has found time for not one *profession* but three— photography, music, and writing—and has reached the top in all three.

UNDERSTANDING THE STORY

>>>> *Circle the letter next to each correct statement.*

1. The main idea of this story is that
   a. Parks won a scholarship that allowed him to study photography.
   b. Parks is a master of the documentary.
   c. Parks is a person of many accomplishments.

2. From this story, you can conclude that
   a. Parks's wide experience is helpful to him in his work.
   b. Parks will win an Academy Award for one of his documentaries.
   c. Parks is ruining his health by trying to do too much.

MAKE AN ALPHABETICAL LIST

>>>> *Here are the ten vocabulary words in this lesson. Write them in alphabetical order in the spaces below.*

notable	pursuing	enthusiastic	profession	scholarship
considerable	documentary	segregation	assignments	plight

1. _____
2. _____
3. _____
4. _____
5. _____

6. _____
7. _____
8. _____
9. _____
10. _____

WHAT DO THE WORDS MEAN?

>>>> *Following are some meanings, or definitions, for the ten vocabulary words in this lesson. Write the words next to their definitions.*

1. _____ an occupation requiring an education

2. _____ striving for

3. _____ money given to help a student pay for studies

4. _____ definite tasks or jobs to be done; specific works to be accomplished

5. _____ eagerly interested

6. _____ a factual presentation of a scene, place, or condition of life in writing or on film

7. _____ a condition or state, usually bad

8. _____ worthy of notice; remarkable

9. _____ not a little; much

10. _____ separation from others; setting individuals or groups apart from society

COMPLETE THE SENTENCES

>>>> *Use the vocabulary words in this lesson to complete the following sentences. Use each word only once.*

scholarship	pursuing	considerable	profession	assignments
enthusiastic	plight	segregation	notable	documentary

1. The _____ *The Learning Tree* is based on Gordon Parks's life.

2. To succeed in one _____ is good, but to succeed in three is exceptional.

3. You are happy to accept all sorts of _____ if you are a young photographer just breaking into the field.

4. Among Parks's _____ achievements is a documentary about a boy in Brazil.

5. The _____ of the poor has always interested Parks.

6. The critics were most _____ in their reviews of Parks's latest film.

7. If he had not received a _____, he could not have finished school.

8. As director, he spent _____ time interviewing actors before picking a lead.

9. Parks made people more aware of the problem of _____.

10. Parks is a person who believes in _____ his interests.

USE YOUR OWN WORDS

>>>> *Look at the picture. What words come into your mind other than the ten vocabulary words used in this lesson? Write them on the lines below. To help you get started, here are two good words:*

1. _____ moustache _____
2. _____ camera _____
3. _____
4. _____
5. _____
6. _____
7. _____
8. _____
9. _____
10. _____

FIND THE SUBJECTS AND PREDICATES

>>>> The **subject** of a sentence names the person, place, or thing that is spoken about. The **predicate** of a sentence is what is said about the subject. For example:

> The small boy went to the football game.

>>>> *The small boy* is the subject (the person the sentence is talking about). *Went to the football game* is the predicate of the sentence (because it tells what the small boy did).

>>>> *In the following sentences, draw one line under the subject of the sentence and two lines under the predicate of the sentence.*

1. I hungered for learning.

2. A scholarship allowed me to finish my education.

3. Two assignments impressed me.

4. The other was a documentary that I wrote.

5. My most notable achievement was writing the book.

COMPLETE THE STORY

>>>> Here are the ten vocabulary words for this lesson:

considerable	profession	assignments	scholarship	enthusiastic
documentary	notable	plight	pursuing	segregation

>>>> *There are six blank spaces in the story below. Four vocabulary words have already been used in the story. They are underlined. Use the other six words to fill in the blanks.*

Gordon Parks has had <u>considerable</u> success in not one _____ but three. However, his most _____ <u>assignments</u> have been on one subject—people.

Parks's _____ love of humanity is shown in many of the picture stories he has photographed over the years. One particular <u>documentary</u> about a Brazilian boy shows what it is like to be poor.

His novel *The Learning Tree*, which is based on his own life, shows the _____ of people who must live under <u>segregation</u>.

It all started for Parks when he was awarded a _____ many years ago. Since then, he has been _____ one special goal—to show how people live. Hopefully, his work will teach us to understand each other better too.

Learn More About Photography

>>>> *On a separate sheet of paper or in your notebook or journal, complete one or more of the activities below.*

Building Language

In a magazine, find a photograph that moves you. If you can, describe the photograph in a language other than English. Write the words in your native language that describe it. Now use English to describe the same picture. How are your descriptions the same or different?

Learning Across the Curriculum

Matthew Brady was one of the first important American photographers. Find out something about Matthew Brady. Look at his photographs of the Civil War. Now find some drawings that illustrate the Revolutionary War. How can looking at photographs change what people think about a war?

Broadening Your Understanding

Cut out photographs and captions from the news section of a newspaper. Then look critically at the photographs. Which photographs do you think work the best? Why? Which photographs do not impress you? What do you think is the most important quality for a news photograph to have? Write a paragraph expressing your ideas.

12 RICH AND AMOS

Horatio Alger, a famous writer of the last century, wrote about rags-to-riches heroes. His characters always became *prosperous* by honest and hard work. So it seems *appropriate* that Wally Amos should be one of the winners of the Horatio Alger Association awards.

Amos is a real Horatio Alger hero. As a young man in Tallahassee, Florida, he was so poor that he had to walk 6 miles to learn a *trade.* He couldn't afford the bus *fare.*

Then Amos started his own business and sold about $12 million worth of his Famous Amos cookies every year. He was the first to open stores that sold only his cookies. You could buy his *products* all over the United States and even in Asia. Wally Amos sold his business in 1985. He now lives in Hawaii and has a new cookie business. He is *definitely* a rags-to-riches hero!

It isn't all hard work, though. Amos is active in *literacy* programs. He strongly supports *efforts* to teach people to read and write. He also wants people to enjoy themselves. "It's okay to have fun while doing important work," he says. He proved his point during a recent meeting to promote literacy by playing "California, Here I Come" on a kazoo.

Perhaps the secret of his success is that Amos enjoys his job. *Authorities* say that successful people like what they do. Amos certainly likes what he does, and so do lots of other people. Amos received one of the first presidential awards for new business excellence from a famous *fan,* Ronald Reagan.

UNDERSTANDING THE STORY

 Circle the letter next to each correct statement.

1. The main purpose of this story is to
   a. tell how to make chocolate chip cookies.
   b. tell about a successful and hard-working man.
   c. explain how to win awards for hard work.

2. From this story, you can conclude that
   a. Amos did not like a regular job.
   b. chocolate chip cookies are easy to make.
   c. people liked Amos's chocolate chip cookies.

MAKE AN ALPHABETICAL LIST

>>>> *Here are the ten vocabulary words in this lesson. Write them in alphabetical order in the spaces below.*

prosperous	definitely	appropriate	literacy	trade
efforts	fare	authorities	products	fan

1. _____

2. _____

3. _____

4. _____

5. _____

6. _____

7. _____

8. _____

9. _____

10. _____

WHAT DO THE WORDS MEAN?

>>>> *Following are some meanings, or definitions, for the ten vocabulary words in this lesson. Write the words next to their definitions.*

1. _____ attempts

2. _____ proper

3. _____ a job; a skill

4. _____ absolutely

5. _____ manufactured items

6. _____ specialists

7. _____ the cost of a ticket

8. _____ an enthusiastic supporter

9. _____ the ability to read and write

10. _____ successful

COMPLETE THE SENTENCES

>>>> *Use the vocabulary words in this lesson to complete the following sentences. Use each word only once.*

prosperous	definitely	appropriate	literacy	trade
efforts	fare	authorities	products	fan

1. Wally Amos _____ had a better idea for cookies.

2. He was too poor to pay his bus _____ as a young man.

3. He learned a _____ in Tallahassee, but it wasn't baking cookies.

4. Through his own _____, Amos became a wealthy businessman.

5. His baked _____ are sold in the United States and Asia.

6. Amos believes that you can have fun even as you become _____.

7. It seems _____ that Amos should tell other people to learn how to have fun.

8. Many _____ have studied how Amos became a success.

9. Amos also gives time to _____ programs.

10. He can even count a President of the United States as a _____.

USE YOUR OWN WORDS

>>>> *Look at the picture. What words come into your mind other than the ten vocabulary words used in this lesson? Write them on the lines below. To help you get started, here are two good words:*

1. _____ smiles _____
2. _____ hats _____
3. _____
4. _____
5. _____
6. _____
7. _____
8. _____
9. _____
10. _____

71

MAKE POSSESSIVE WORDS

>>>> The possessive of a word shows that something belongs to it. For example, Wally has a plan; it is *Wally's* plan. To make a possessive of a word that doesn't end in *s*, add an apostrophe and an *s* to the word, such as *cookie's* price or *store's* shelf. To make a possessive of a word that does end in *s*, add an apostrophe (') after the *s*, such as *cookies'* chips or *stores'* sales. (A proper name ending in *s* is an exception: *Amos's* cookies.)

>>>> *Here are ten words from the story. In the blank space next to each word, write the correct possessive of each word.*

1. Alger _____

2. heroes _____

3. people _____

4. meeting _____

5. kazoo _____

6. United States _____

7. awards _____

8. Reagan _____

9. winners _____

10. rags _____

COMPLETE THE STORY

>>>> Here are the ten vocabulary words for this lesson:

prosperous	definitely	appropriate	literacy	trade
efforts	fare	authorities	products	fan

>>>> *There are six blank spaces in the story below. Four vocabulary words have already been used in the story. They are underlined. Use the other six words to fill in the blanks.*

A _____ businessperson must be ready for hard work. The main idea is to sell the _____ of the business. Success takes great <u>efforts</u> each day. It is _____ not easy. You may become discouraged as you give your bus _____ to the driver again and again every day. Yet you should never give up hope!

There are some things you can do to prepare for success. Reading and writing are important, so work on your _____ skills. The <u>authorities</u> who study business success also say you must be prepared to do things yourself. That is _____ after all, if you want to know how your business works. You can go to <u>trade</u> school to learn some useful business techniques. Read letters from your <u>fans</u>. They will tell you what to change and what to leave alone.

Learn More About Business

>>>> *On a separate sheet of paper or in your notebook or journal, complete one or more of the activities below.*

Working Together

Imagine your group is going to start a business. Decide what your business will be. Now write a business plan for your business. Have people research how much your product will cost to produce and market, how much you think you can sell it for, how many you think you can sell, and what you think your profits will be. Have someone plan an advertising campaign. Someone else can design the packaging and plan where to sell your product. Put your plans together and present them to the class.

Broadening Your Understanding

You are in charge of creating an advertising campaign to sell Wally Amos's cookies. Think about the product and how you can interest people in it. Think of how you can show that Famous Amos cookies are better than other chocolate chip cookies. Think of who you want to buy these cookies. Now write a newspaper or magazine ad that will sell Famous Amos cookies.

Learning Across the Curriculum

What kind of business intrigues you? Research a business you are interested in. Find out what the people do in that business. Write a summary of what you found out and whether you still think you would like to be involved in that business.

13 A DAYTIME STAR

Oprah Winfrey, the popular actress and talk-show host, **embodies** an important message. That message is this: You can be born poor, black, and female and make it to the top.

Winfrey spent her first six years with her grandmother, who she says could "whip me for days and never get tired." Winfrey says her mother worked hard and wanted the best for her but did not know how to achieve these aims.

As a teenager, Winfrey rebelled and got into trouble. She went to Tennessee to live with her father, a strict **disciplinarian** who encouraged her to read a book every week. Years later, in 1971, Oprah became Miss Black Tennessee. In 1976, she joined the ABC **affiliate** WJZ-TV in Baltimore as conewsperson. But it was her film acting that brought her national fame. Winfrey won an Academy Award nomination for her role as Sophia in the film of Alice Walker's novel *The Color Purple*. "Luck," says Winfrey, "is a **matter** of preparation. I've been **blessed**—but I create the blessings."

Today Oprah Winfrey's name is **synonymous** with daytime talk shows. She has won six Emmys for best talk-show host. She deals with important social issues on her show. Many people think of her as an **activist.** She is concerned about **racial** problems in America. Each week, millions of Americans eagerly **await** her shows to hear about the **issues** she presents.

UNDERSTANDING THE STORY

 Circle the letter next to each correct statement.

1. The main idea of this story is that
   a. Oprah Winfrey was nominated for an Academy Award.
   b. despite a poor start in life, Winfrey has become very accomplished.
   c. Oprah Winfrey is upset by racial inequality.

2. From the story, you can conclude that
   a. people can overcome difficulties to accomplish what they want.
   b. Winfrey will start a new career in South Africa.
   c. Winfrey was unsuccessful as an actress.

MAKE AN ALPHABETICAL LIST

>>>> *Here are the ten vocabulary words in this lesson. Write them in alphabetical order in the spaces below.*

issues	embodies	affiliate	matter	await
blessed	racial	synonymous	activist	disciplinarian

1. _____
2. _____
3. _____
4. _____
5. _____

6. _____
7. _____
8. _____
9. _____
10. _____

WHAT DO THE WORDS MEAN?

>>>> *Following are some meanings, or definitions, for the ten vocabulary words in this lesson. Write the words next to their definitions.*

1. _____ alike in meaning or significance

2. _____ a person who believes in strict training

3. _____ a person who publicly supports a cause

4. _____ a person or an organization usually connected to a larger organization

5. _____ a real thing; content rather than manner or style

6. _____ of or having to do with race or origins

7. _____ topics or problems under discussion

8. _____ given great happiness

9. _____ to wait for; to expect

10. _____ represents in real or definite form

COMPLETE THE SENTENCES

>>>> *Use the vocabulary words in this lesson to complete the following sentences. Use each word only once.*

issues	embodies	affiliate	matter	await
blessed	racial	synonymous	activist	disciplinarian

1. Winfrey has become an _____ for important causes.

2. Her program focuses on important _____ of the day.

3. Winfrey really _____ a rags-to-riches story.

4. Winfrey's father was a _____ who encouraged education.

5. The TV station Winfrey worked for was an _____ of a larger station.

6. Winfrey admits that much of her success is a _____ of preparation.

7. Winfrey is very aware of the _____ problems in our society.

8. For many viewers, the name Oprah Winfrey is _____ with television talk shows.

9. Winfrey says she is _____ because she helps bring about her luck.

10. Winfrey's fans _____ her show every day to see who her guests will be.

USE YOUR OWN WORDS

>>>> *Look at the picture. What words come into your mind other than the ten vocabulary words used in this lesson? Write them on the lines below. To help you get started, here are two good words:*

1. _____ happy _____
2. _____ smiling _____
3. _____
4. _____
5. _____
6. _____
7. _____
8. _____
9. _____
10. _____

DO THE CROSSWORD PUZZLE

>>>> *In a crossword puzzle, there is a group of boxes, some with numbers in them. There are also two columns of words, or definitions, one for "across" and the other for "down." Do the puzzle. Each of the words in the puzzle will be one of the vocabulary words in this lesson.*

Across

3. having to do with race
5. strict person
9. public supporter of causes
10. problems

Down

1. a real thing; content
2. makes real
4. to have great happiness
6. alike in meaning
7. member
8. to look forward to

COMPLETE THE STORY

>>>> Here are the ten vocabulary words for this lesson:

issue	embodies	affiliate	matter	await
blessed	racial	synonymous	activist	disciplinarian

>>>> *There are six blank spaces in the story below. Four vocabulary words have already been used in the story. They are underlined. Use the other six words to fill in the blanks.*

Oprah Winfrey is a successful TV talk-show host. She started working at an ABC <u>affiliate</u> station in Baltimore. Winfrey considers herself _____, but she knows that much of her luck has been a _____ of work. She has become an <u>activist</u>. Every show focuses on an important social <u>issue</u>. For example, she is very aware of <u>racial</u> discrimination. Although her father was a _____, Winfrey had a difficult childhood. Through work and determination, her name has become _____ with daytime TV. Winfrey _____ the American idea that, people can succeed against great odds. Her fans _____ her next achievement.

78

Learn More About Talk Shows

>>>> *On a separate sheet of paper or in your notebook or journal, complete one or more of the activities below.*

Building Language

Watch a talk show on television. List the words and phrases you do not understand. Either look up the words and phrases or ask a friend what each one means. Use each word or phrase in a sentence.

Broadening Your Understanding

Imagine you are in charge of finding topics and guests for Oprah Winfrey's talk show. Plan a week's worth of topics and the guests you would want for each one. Then explain why you chose what you did.

Extending Your Reading

Read one of these books about television. Then plan your own talk show. What staff will you need to help put the show on the air? Consider everything from the people who plan the show to the people who help produce the show.

Careers in Television, by Howard Blumenthal
Series TV, by Malka Drucker and Elizabeth James
Make Your Own . . .Videos, Commercials, Radio Shows, by the Fun Group

14 FREEDOM FIGHTER

The stage was bare except for a piano, a bench, and one empty chair. Although the audience sat quietly, a feeling of excitement filled the room. Something great was about to happen.

The houselights dimmed. A short, stout, bald man, carrying a `cello` and a bow in one hand, walked slowly to the chair. He turned and faced the audience. The theater erupted with cheers and wild applause. Pablo Casals, the world's greatest cellist, was about to bring the room alive with his music.

Born in Spain, Casals became a freedom fighter during the Spanish civil war. But when this revolution was over, General Franco had won control over the Spanish people. He formed a `dictatorship` that `restricted` the freedom of all the Spanish people. Those who disagreed with him were shot, put in prison, or `banished` from their homeland. Casals became a sworn enemy of this `tyranny.`

Casals left his native land in `protest` against Franco's government. But he continued to fight for freedom. He `devoted` his life to helping the people of Spain. Casals organized benefit concerts in France and Puerto Rico. These music `festivals` raised money to help Spanish exiles. Casals was `motivated` by his love of freedom.

Casals played from his heart. The music from his stringed instrument was `flawless`—smooth and moving. When he died in Puerto Rico at the age of 96, Casals left behind an example for all freedom-loving people of the world.

UNDERSTANDING THE STORY

>>>> *Circle the letter next to each correct statement.*

1. Another good title for this story might be:
   a. "The World's Greatest Musician."
   b. "The Dangers of Dictatorship."
   c. "An Artist Against Tyranny."

2. From this story, you can conclude that
   a. people should not mix art with politics.
   b. Pablo Casals took as much pride in his political beliefs as in his music.
   c. Pablo Casals gave up his music to fight for freedom.

MAKE AN ALPHABETICAL LIST

>>>> *Here are the ten vocabulary words in this lesson. Write them in alphabetical order in the spaces below.*

cello	festivals	restricted	motivated	protest
devoted	dictatorship	tyranny	banished	flawless

1. _____
2. _____
3. _____
4. _____
5. _____

6. _____
7. _____
8. _____
9. _____
10. _____

WHAT DO THE WORDS MEAN?

>>>> *Following are some meanings, or definitions, for the ten vocabulary words in this lesson. Write the words next to their definitions.*

1. _____ stimulated to do something; inspired

2. _____ a musical instrument similar to a violin, but much larger, that is played in a sitting position

3. _____ limited in freedom or use

4. _____ celebrations

5. _____ the rule by one person or group that must be obeyed; power held by a few through force

6. _____ strong objection; opposition

7. _____ perfect; without fault

8. _____ forced to leave one's country

9. _____ gave up one's time, money, or efforts for some cause or person

10. _____ the cruel use of power

COMPLETE THE SENTENCES

>>>> *Use the vocabulary words in this lesson to complete the following sentences. Use each word only once.*

devoted	cello	protest	tyranny	flawless
dictatorship	banished	motivated	festivals	restricted

1. In _____ against the government of Franco, Pablo Casals left Spain.

2. Casals continued to fight against _____.

3. When freedom of speech is _____, it is difficult for people to express their true feelings.

4. Those who protest are often _____ from their homeland.

5. The _____ of a dictator causes some people to want to leave.

6. Casals _____ most of his time to organizing benefit concerts.

7. The music _____ he organized were successful.

8. When people talk about masters of the _____, Pablo Casals is the name most often mentioned.

9. Casals was _____ by his deep love of music to be a great cellist.

10. His playing was described as _____ and inspiring.

USE YOUR OWN WORDS

>>>> *Look at the picture. What words come into your mind other than the ten vocabulary words used in this lesson? Write them on the lines below. To help you get started, here are two good words:*

1. _____ bow _____
2. _____ strings _____
3. _____
4. _____
5. _____
6. _____
7. _____
8. _____
9. _____
10. _____

MATCH TERMS WITH THEIR MEANINGS

>>>> *Here are some words selected from the field of music. See if you can match the terms with their meanings.*

1. **symphony** _____   **a.** the speed at which music is played.

2. **viola** _____   **b.** a long musical work for an orchestra

3. **tempo** _____   **c.** when music gradually becomes louder

4. **tympanist** _____   **d.** a member of the violin family that is between the violin and cello in size

5. **crescendo** _____   **e.** the orchestra member who plays the kettledrums

COMPLETE THE STORY

>>>> Here are the ten vocabulary words for this lesson:

devoted	restricted	dictatorship	cello	motivated
festivals	banished	tyranny	flawless	protest

>>>> *There are six blank spaces in the story below. Four vocabulary words have already been used in the story. They are underlined. Use the other six words to fill in the blanks.*

Pablo Casals, the great _____ master, left his native Spain in _____ against General Franco's <u>dictatorship.</u>

Casals was <u>devoted</u> to peace and to helping the people of Spain. He organized music _____ to raise money for those Spaniards who had been _____ from their homeland. In so doing, he <u>motivated</u> other people to demonstrate against _____.

Through his _____ music, Casals took a courageous stand against any government that <u>restricted</u> the freedom of its citizens.

Learn More About Music

>>>> *On a separate sheet of paper or in your notebook or journal, complete one or more of the activities below.*

Learning Across the Curriculum

Music can stir people to action. Look in books about folk songs to find a song that was popular during a war or another time of stress for the United States. Listen to a recording of the song, if possible. Read the words. Then describe what effect you think the song might have had on people's opinions.

Broadening Your Understanding

Find out more about the Spanish civil war in which Pablo Casals was involved. What did a freedom fighter do during the Spanish civil war? When was this revolution? What was the result of this war? In what way do you think being a freedom fighter may have influenced the way Casals performed his music?

Extending Your Reading

Casals organized benefit concerts to help raise money for Spanish exiles. In this country, musicians have held concerts to benefit various causes. Read one of the books below about the Live Aid concerts. Then plan your own benefit concert. What cause would you benefit? What musicians would you hope to get involved?

Live Aid, by Susan Clinton
Bob Geldof: The Pop Star Who Raised $170 Million for Famine Relief in Ethiopia, by Vallance D'Ar Adrian

15 MOUNTAIN OF FIRE

Our bus wound its way slowly up the steep Italian mountainside through olive groves, fruit orchards, and thick forests. At 6,500 feet, we reached the end of the road. From this point on, we traveled by cable car.

The view below us was unusually beautiful. As far as we could see, the mountain was covered with sand, hardened *lava,* and ashes. When we reached the end of the cable-car run, we climbed out. It was cold, but we would have to walk the rest of the way.

Our guide said, "This is a very dangerous climb. If anyone falls from the top, he or she will *vanish* forever. Therefore, I must *enforce* safety rules."

He went on. "An active *volcano* is a weak place in the earth's crust. Extremely hot gases *exert* tremendous pressure, force their way to the top, and escape from the earth's *interior.* It can *erupt* at any time, sending rivers of molten lava down the mountainside. A few years ago, the village of Santa d'Alfio was destroyed during an eruption. But the brave people rebuilt their town."

We carefully made our way farther up the mountain. When we reached the top, we *gingerly* stepped to the edge of the fiery crater, where ropes kept us from getting too close. Clouds of smoke, steam, and black dust rose into the air, causing our eyes to *smart.* Breathing was difficult. We had reached the *peak* of Europe's most active volcano—Mount Etna.

UNDERSTANDING THE STORY

 Circle the letter next to each correct statement.

1. The main purpose of this story is to
   a. tell about the violent history of Mount Etna.
   b. tell us never to go near a volcano.
   c. describe the adventure of a group of people climbing a volcano.

2. If a volcano is said to be active, you can assume that
   a. no one is living in the area around it.
   b. it can erupt at any time.
   c. it will have a major eruption in the next month.

MAKE AN ALPHABETICAL LIST

>>>> *Here are the ten vocabulary words in this lesson. Write them in alphabetical order in the spaces below.*

enforce	lava	peak	exert	interior
smart	gingerly	vanish	erupt	volcano

1. _____   6. _____

2. _____   7. _____

3. _____   8. _____

4. _____   9. _____

5. _____   10. _____

WHAT DO THE WORDS MEAN?

>>>> *Following are some meanings, or definitions, for the ten vocabulary words in this lesson. Write the words next to their definitions.*

1. _____ to feel a sharp pain; to sting

2. _____ to disappear

3. _____ to make someone do something; to compel

4. _____ to explode; to burst forth

5. _____ inside; inner part

6. _____ melted rock that comes from a volcano

7. _____ the top; the highest point

8. _____ to apply; to use fully

9. _____ very carefully

10. _____ a mountain with a cuplike crater that throws out hot melted rock and steam

COMPLETE THE SENTENCES

>>>> *Use the vocabulary words in this lesson to complete the following sentences. Use each word only once.*

smart	erupt	interior	exert	lava
peak	enforce	vanish	volcano	gingerly

1. Our eyes began to _____ from the Volcano's smoke.

2. The guide said that an active volcano may _____ at any time.

3. We stepped _____ over the broken pieces of rock.

4. After a long uphill climb, we finally reached the _____ of the volcano.

5. Volcanoes result when forces beneath the earth's surface _____ pressure.

6. Mount Etna is considered to be the most active _____ in Europe.

7. We wanted to peer into the _____ of the volcano.

8. Police _____ laws that prohibit people from getting close to the volcano.

9. The _____ streaming down the mountainside looked like a river of fire.

10. Whole villages have been known to _____ from sight after an eruption.

USE YOUR OWN WORDS

>>>> *Look at the picture. What words come into your mind other than the ten vocabulary words used in this lesson? Write them on the lines below. To help you get started, here are two good words:*

1. _____ smoke _____
2. _____ mountain _____
3. _____
4. _____
5. _____
6. _____
7. _____
8. _____
9. _____
10. _____

>>>> A **synonym** is a word that means the same, or nearly the same, as another word. An **antonym** is a word that means the opposite of another word.

>>>> *There are six vocabulary words listed below. To the right of each is either a synonym or an antonym. On the line beside each pair of words, write S for synonyms or A for antonyms.*

1. **vanish**	appear	1. _____
2. **peak**	bottom	2. _____
3. **smart**	sting	3. _____
4. **gingerly**	carelessly	4. _____
5. **interior**	exterior	5. _____
6. **erupt**	gush	6. _____

COMPLETE THE STORY

>>>> Here are the ten vocabulary words for this lesson:

enforce	lava	peak	exert	interior
smart	gingerly	vanish	erupt	volcano

>>>> *There are six blank spaces in the story below. Four vocabulary words have already been used in the story. They are underlined. Use the other six words to fill in the blanks.*

A <u>volcano</u> is active when the <u>interior</u> gases can push their way upward to the surface. An active volcano can produce rivers of molten _____.

A volcano is inactive if it cannot _____.

One must _____ caution when visiting an active volcano. You must step <u>gingerly</u> around the crater. Careless people have been known to fall into the crater and suddenly _____. To prevent such accidents, officials _____ strict safety regulations.

While on the <u>peak</u> of an active volcano, a visitor might have to use a handkerchief to cover his or her mouth. The gases can burn your throat and make your eyes _____.

Learn More About Volcanoes

>>>> *On a separate sheet of paper or in your notebook or journal, complete one or more of the activities below.*

Building Language

Several of the vocabulary words in this selection can be confusing because they seem to have more than one meaning. Write definitions for *smart* and *peak*. Then look up and write another meaning for the words. Now write a definition for the word *gingerly*. How could this word be easily misunderstood?

Working Together

Create a three-dimensional map of the world with the major volcanoes labeled. Divide up the tasks so that some people are researching volcanoes in different continents and others are constructing and labeling the map. Display your map in the classroom.

Learning Across the Curriculum

The eruption of Mount Vesuvius in Pompeii, Italy, in A.D. 79 was remarkable for what was preserved when the lava poured over the unsuspecting town. Look up information about the destruction of Pompeii. Pretend you are a news reporter on the scene, broadcasting the news of what has happened. Write a radio report, explaining the eruption and its results.

16 GOING ONCE, TWICE...

The proud colt steps into the auction ring. He is turned in slow circles. His brown coat has been brushed to a high $\boxed{gloss.}$ The horse breeders study the thoroughbred. The $\boxed{auctioneer}$ begins his $\boxed{chant.}$ The bidding starts at a quarter of a million dollars. Quickly, the price rises to a half a million. Breeders from many different countries $\boxed{compete}$ against one another. The final bid is one and a half million dollars. The colt now has an owner.

The Keeneland race course in Kentucky is the scene of many multimillion-dollar bids. Buyers look for a horse with an excellent $\boxed{pedigree.}$ In this way, they improve the bloodlines of their stables.

One likely colt sold for $4.5 million. He was the son of the great racehorse Nijinsky II. Even more amazing are the prices paid when groups of people buy horses $\boxed{jointly.}$ In shared ownership, each partner has the right to breed the horse once a year. Conquistador Cielo was bought this way for over $36 million! His owners hoped he would $\boxed{sire}$ many winners.

Splendid $\boxed{banquets}$ are given to attract horse buyers to auctions. These banquets have become a $\boxed{permanent}$ part of the auction scene. In addition to enjoying themselves, buyers study each horse carefully. Future owners $\boxed{investigate}$ every aspect of a horse before bidding. They even use computers to trace bloodlines. Buying racehorses is no small matter. Going once, going twice. . . sold!

UNDERSTANDING THE STORY

 Circle the letter next to each correct statement.

1. The main idea of this story is that
   a. huge sums of money are spent at auctions on horses that can win races and sire other great racehorses.
   b. buying horses jointly helps breeders to afford them.
   c. the use of computers in tracing bloodlines has revolutionized the business of buying racehorses.

2. From this story, you can conclude that
   a. thoroughbreds will soon be inexpensive because there will be so many of them.
   b. the record price paid for Conquistador Cielo will not seem high in years to come.
   c. a flowing mane and large front hooves are signs of a great racehorse.

MAKE AN ALPHABETICAL LIST

>>>> *Here are the ten vocabulary words in this lesson. Write them in alphabetical order in the spaces below.*

gloss	permanent	auctioneer	jointly	compete
sire	investigate	banquets	chant	pedigree

1. _____
2. _____
3. _____
4. _____
5. _____

6. _____
7. _____
8. _____
9. _____
10. _____

WHAT DO THE WORDS MEAN?

>>>> *Following are some meanings, or definitions, for the ten vocabulary words in this lesson. Write the words next to their definitions.*

1. _____ to oppose; to try for the same thing

2. _____ the agent in charge of selling at an auction

3. _____ formal meals; lavish feasts

4. _____ high polish; shine

5. _____ to study; to look into carefully

6. _____ together; in partnership

7. _____ to be the father of

8. _____ rapid and rhythmic speaking

9. _____ the record of an animal's ancestors, especially with respect to purity of breed

10. _____ lasting; continuing

COMPLETE THE SENTENCES

>>>> *Use the vocabulary words in this lesson to complete the following sentences. Use each word only once.*

chant	sire	auctioneer	jointly	permanent
banquets	gloss	compete	pedigree	investigate

1. When the _____ banged her gavel on the stand, it meant that the sale was over.

2. The auctioneer's _____ sounded like a tuneless song.

3. Great _____ are served at horse auctions in order to attract bidders.

4. The tension in the room rises as breeders _____ against one another.

5. A breeder will _____ a horse's health and bloodline before bidding.

6. A breeder wants to be sure that the _____ of a horse is a good one.

7. Buyers sometimes get together and purchase a horse _____.

8. A good diet and daily brushing give a high _____ to a horse's coat.

9. A horse can _____ many young, but only a few may grow up to be great racehorses.

10. A racehorse often becomes a be _____ part of a well-known stable.

USE YOUR OWN WORDS

>>>> *Look at the picture. What words come into your mind other than the ten vocabulary words used in this lesson? Write them on the lines below. To help you get started, here are two good words:*

1. _____ bidder _____
2. _____ ropes _____
3. _____
4. _____
5. _____
6. _____
7. _____
8. _____
9. _____
10. _____

>>>> A **synonym** is a word that means the same, or nearly the same, as another word. An **antonym** is a word that means the opposite of another word.

>>>> *There are six vocabulary words listed below. To the right of each is either a synonym or an antonym. On the line beside each pair of words, write S for synonyms or A for antonyms.*

1. **compete**	cooperate	1. _____
2. **pedigree**	history	2. _____
3. **jointly**	separately	3. _____
4. **banquets**	fancy meals	4. _____
5. **permanent**	temporary	5. _____
6. **auctioneer**	buyer	6. _____

COMPLETE THE STORY

>>>> Here are the ten vocabulary words for this lesson:

compete	sire	auctioneer	banquets	pedigree
investigate	gloss	permanent	chant	jointly

>>>> *There are six blank spaces in the story below. Four vocabulary words have already been used in the story. They are underlined. Use the other six words to fill in the blanks.*

Horse auctions are exciting. The horses' coats are brushed to a high
_____. The _____ leads the show with a <u>chant</u> that
is rapid and singsong. Tasty and attractive _____ are served to bring
in buyers. These feasts have become a _____ part of the scene.

You can be sure that buyers _____ every detail of a horse's
history before bidding. The <u>pedigree</u> is studied to see how many of the horse's
ancestors were winners. The breeders <u>compete</u> for horses with excellent records.

Because racehorses are so costly, people often decide to buy them
_____ . They hope that the horse will <u>sire</u> many great racers.

Learn More About Racing Horses

>>>> *On a separate sheet of paper or in your notebook or journal, complete one or more of the activities below.*

Learning Across the Curriculum

Owning horses and racing them are very costly. Research the expenses involved in owning, training, and racing a horse. Find out how much the average racehorse costs its owner a year. Give a presentation to your classmates.

Broadening Your Understanding

The Kentucky Derby is one of the most celebrated horse races in the United States. Find out the history of this race and write about it. Include information about some of the most famous winners of the Kentucky Derby. Present your findings to the class.

Extending Your Reading

Writer Dick Francis is a jockey who became a mystery writer. He never forgot his love of horses, and his books are all mysteries about horses. Listed below are several of his books. Read one. Then write a paragraph and discuss whether you think the information in his book about the business of horses is realistic.

Blood Sport
Bolt
Bone Crack
Forfeit
Hot Money

17 THE MYSTERY OF STONEHENGE

One of the most mysterious places in the world is Stonehenge, in England. Here, an ancient *monument* of rock dominates the landscape.

At Stonehenge, there are three circles of huge stones, one inside the other. Each stone is *approximately* 14 feet high and weighs about 28 tons. Next to each stone in the outer circle is a shallow pit. Inside the smallest circle are stones shaped like horseshoes. At the center is a fallen stone that could have been an altar.

The rocks of Stonehenge still *perplex* scientists today. Carvings on the stones *denote* that they were put there about 4,000 years ago. Some *pundits* believe Stonehenge was built for sun worship. Others say it might have been built by *pagan* priests as a temple or a *burial* place. *Cremated* bodies have been found in the pits. Still other people believe Stonehenge was an *astronomical* calendar used to tell the seasons of the year and *eclipses* of the sun and moon.

There are also many questions about how the stones got there. Were the stones hauled by hand or floated in on rafts? How were they raised to an upright position?

No one has been able to answer these questions. Stonehenge remains one of the world's biggest mysteries.

UNDERSTANDING THE STORY

 Circle the letter next to each correct statement.

1. The main idea of this story is that
   a. most scientists believe Stonehenge was an astronomical calendar.
   b. the origin of Stonehenge remains a mystery, although many theories have been suggested.
   c. Stonehenge is considered one of the seven wonders of the world.

2. The fact that human ashes were found in the pits at Stonehenge suggests that
   a. the area was used as a battlefield.
   b. the stones are what is left of a larger structure that burned to the ground.
   c. bodies were often cremated before burial in ancient times.

MAKE AN ALPHABETICAL LIST

>>>> *Here are the ten vocabulary words in this lesson. Write them in alphabetical order in the spaces below.*

monument	burial	perplex	approximately	denote
pundits	pagan	cremated	astronomical	eclipses

1. _____
2. _____
3. _____
4. _____
5. _____

6. _____
7. _____
8. _____
9. _____
10. _____

WHAT DO THE WORDS MEAN?

>>>> *Following are some meanings, or definitions, for the ten vocabulary words in this lesson. Write the words next to their definitions.*

1. _____ burned to ashes

2. _____ persons who have knowledge of a subject

3. _____ to puzzle; to confuse

4. _____ to show; to point out

5. _____ times when the sun or moon cannot be seen because its light is blocked

6. _____ something from a past age that is believed to have historical importance

7. _____ having to do with astronomy; the study of planets, stars, and other bodies in outer space

8. _____ a follower of a religion with many gods

9. _____ having to do with placing a body in its final resting place

10. _____ nearly; about

>>>> *Use the vocabulary words in this lesson to complete the following sentences. Use each word only once.*

pundits	monument	denote	cremated	eclipses
pagan	astronomical	perplex	burial	approximately

1. A question that continues to _____ scientists is what was the purpose of Stonehenge.

2. Stonehenge may have predicted _____ of the sun and moon.

3. In fact, drawings of ancient _____ calendars show structures that look much like Stonehenge.

4. Priests of a _____ religion may have meant the stones to be a temple.

5. Carvings on the stones _____ that Stonehenge is about 4,000 years old.

6. One of the things that _____ wonder is what the purpose of the Stonehenge monument was.

7. The ashes of _____ bodies were found in the area of the outer circle.

8. Stonehenge may have been a sacred _____ place.

9. A wall of earth _____ 320 feet in diameter surrounds the stones.

10. The _____ of Stonehenge stands as a reminder of ancient people.

USE YOUR OWN WORDS

>>>> *Look at the picture. What words come into your mind other than the ten vocabulary words used in this lesson? Write them on the lines below. To help you get started, here are two good words:*

1. _____ grass _____
2. _____ sky _____
3. _____
4. _____
5. _____
6. _____
7. _____
8. _____
9. _____
10. _____

FIND THE ANALOGIES

>>>> In an **analogy,** similar relationships occur between words that are different. For example, *pig* is to *hog* as *car* is to *automobile*. The relationship is that the words mean the same. Here's another analogy: *noisy* is to *quiet* as *short* is to *tall*. In this relationship, the words have opposite meanings.

>>>> *See if you can complete the following analogies. Circle the correct word or words.*

1. **Denote** is to **point out** as **dominates** is to

a. rises above   **b.** towers over   **c.** grows up   **d.** climbs over

2. **Stonehenge** is to **monument** as **eclipses** is to

a. sunlight   **b.** planets   **c.** heavenly bodies   **d.** blackouts

3. **Mysterious** is to **strange** as **ancient** is to

a. approximately   **b.** old   **c.** pagan   **d.** astronomical

4. **Perplex** is to **confuse** as **cremated** is to

a. boiled   **b.** buried   **c.** cooled   **d.** burned

5. **Erected** is to **destroyed** as **upright** is to

a. fallen   **b.** vertical   **c.** approximately   **d.** astronomical

COMPLETE THE STORY

>>>> Here are the ten vocabulary words for this lesson:

monument	approximately	cremated	denote	pundits
pagan	perplex	astronomical	eclipses	burial

>>>> *There are six blank spaces in the story below. Four vocabulary words have already been used in the story. They are underlined. Use the other six words to fill in the blanks.*

The rock _____ of Stonehenge dominated the Salisbury Plains in England.

The rocks of Stonehenge still _____ pundits today. Some believe Stonehenge was an _____ calendar that predicted eclipses of the moon and sun. Others think it was used as a _____ temple or _____ place. Cremated bodies have been found there. Carvings denote the stones were erected _____ 4,000 years ago.

Despite modern science, Stonehenge remains a mystery.

102

Learn More About Ancient Science

> **>>>>** *On a separate sheet of paper or in your notebook or journal, complete one or more of the activities below.*

Working Together

Scientists have many theories about why Stonehenge was built. Research the theories and assign one student to find out about each theory. Then have a debate with each person defending the theory he or she researched. Have the class vote to decide which person was most persuasive.

Learning Across the Curriculum

The Mayans, the Egyptians, and many other ancient civilizations had ways of telling time. Research the way one of these peoples kept track of time and write a paragraph explaining it. If you can, build a model and demonstrate it for the class.

Broadening Your Understanding

Find out more about the Druids, who scientists believe built Stonehenge. Write a description of what they believed and how they lived. Also write something about their history.

18 TRUE TO HER OWN VISION

As a child, Georgia O'Keeffe showed a talent for art. She enjoyed expressing her feelings in images. One day Alfred Stieglitz, a famous critic and photographer, saw some of her work. He liked it so much that he **exhibited** her paintings in a one-woman show. That show, in 1916, made O'Keeffe's reputation as an important artist. She married Stieglitz. He encouraged her to express her vision in her own way. Giant flowers, barns, sharply drawn buildings, and bright landscapes emerged from her brush.

O'Keeffe's paintings display freshness and **vigor.** Her works vary a great deal. This makes **classification** of her style difficult. Some critics label her a **romanticist.** Others say she is a realist.

O'Keeffe moved to New Mexico. **Rural,** or country, scenes appealed to her. She found them **preferable** to urban, or city, subjects. Canyons, adobe houses, and **vast** landscapes of the Southwest appeared in many of her paintings. She painted a skull **ornamented** with a bright flower to show life and death together. Some of her paintings became dreamy and **impressionistic.** Others were sharp and clear.

Georgia O'Keeffe died in 1986. Her popularity continues to increase. In 1989, an O'Keeffe **retrospective** was exhibited in major museums across the country, an honor reserved for the best artists.

UNDERSTANDING THE STORY

 Circle the letter next to each correct statement.

1. The main idea of this story is that
   a. Georgia O'Keeffe was married to famous critic and photographer Alfred Stieglitz.
   b. the variety and vigor of O'Keeffe's painting place her among the best artists of this century.
   c. O'Keeffe's paintings usually show rural rather than urban scenes.

2. From this story, you can conclude that
   a. the American Academy of Arts and Letters elected O'Keeffe a member on the basis of her skull paintings alone.
   b. O'Keeffe's career as a painter ended when she married Stieglitz.
   c. O'Keeffe's paintings present her personal vision of life in a way that appeals to many people.

MAKE AN ALPHABETICAL LIST

>>>> *Here are the ten vocabulary words in this lesson. Write them in alphabetical order in the spaces below.*

retrospective	romanticist	exhibited	vast	classification
ornamented	rural	preferable	vigor	impressionistic

1. _____   6. _____

2. _____   7. _____

3. _____   8. _____

4. _____   9. _____

5. _____   10. _____

WHAT DO THE WORDS MEAN?

>>>> *Following are some meanings, or definitions, for the ten vocabulary words in this lesson. Write the words next to their definitions.*

1. _____ putting something into a special group or class

2. _____ displayed; shown to the public

3. _____ something liked better; more desirable

4. _____ huge; spacious

5. _____ strength; vitality

6. _____ an exhibition of the life work of an artist

7. _____ having to do with open country and farming

8. _____ decorated; made more beautiful

9. _____ one who paints people and things as she or he would like them to be rather than as they really are

10. _____ in the style of painting in which the painter tries to catch a momentary glimpse of the subject

COMPLETE THE SENTENCES

>>>> *Use the vocabulary words in this lesson to complete the following sentences. Use each word only once.*

vigor	vast	preferable	classification	impressionistic
romanticist	rural	ornamented	retrospective	exhibited

1. _____ painters try to capture the effect of sunlight in their paintings.

2. _____ of O'Keeffe's work is difficult because different styles appear in it.

3. A _____ is likely to paint landscapes.

4. O'Keeffe was a young, unknown artist when Stieglitz _____ her work.

5. The brush strokes in O'Keeffe's paintings show _____ and spirit.

6. The _____ lands of New Mexico greatly impressed O'Keeffe.

7. Though most of her paintings are simple, some are _____ with brightly colored leaves and flowers.

8. As a child, O'Keeffe found it _____ to express herself in painted images.

9. After her death, many museums displayed the O'Keeffe _____.

10. Houses made of adobe are common in the _____ Southwest.

USE YOUR OWN WORDS

>>>> *Look at the picture. What words come into your mind other than the ten vocabulary words used in this lesson? Write them on the lines below. To help you get started, here are two good words:*

1. _____ work _____
2. _____ shapes _____
3. _____
4. _____
5. _____
6. _____
7. _____
8. _____
9. _____
10. _____

FIND SOME SYNONYMS

>>>> A **synonym** is a word that means the same, or nearly the same, as another word. *Sorrowful* and *sad* are synonyms.

>>>> *The story you read has many interesting words that were not highlighted as vocabulary words. Six of these words are* **vary, scene, critic, express, talent,** *and* **vision.** *Can you think of a synonym for each of these words? Write the synonym in the blank space next to the word.*

1. vary _____

2. scene _____

3. critic _____

4. express _____

5. talent _____

6. vision _____

COMPLETE THE STORY

>>>> Here are the ten vocabulary words for this lesson:

| retrospective | vigor | classification | impressionistic | preferable |
| romanticist | vast | ornamented | rural | exhibited |

>>>> *There are six blank spaces in the story below. Four vocabulary words have already been used in the story. They are underlined. Use the other six words to fill in the blanks.*

When Alfred Stieglitz first _____ the works of Georgia O'Keeffe, the critics were amazed. They found it hard to believe that this artist painted so well in so many styles. <u>Classification</u> of her paintings into one group seemed impossible. Some viewers thought her art was _____ in style. Others said that she was a <u>romanticist</u> in her outlook. She soon achieved a reputation as one of the finest artists in the United States.

O'Keeffe's paintings show a <u>vigor</u> that is inspiring. The dried bones and _____ deserts she paints are often brightly _____ with huge flowers or plants. While urban artists find it <u>preferable</u> to paint city scenes, her works show scenes of _____ life. People could see examples of her different styles in the _____ displayed after her death.

108

Learn More About Artists

>>>> *On a separate sheet of paper or in your notebook or journal, complete one or more of the activities below.*

Appreciating Diversity

Research the work of an artist who painted pictures of the landscape in a region or country in which you are interested. Also look at photographs of this area. How did the artist interpret the landscape?

Learning Across the Curriculum

Georgia O'Keeffe is not the only artist to be inspired by the American Southwest. Find an example of the work of painter R.C. Gorman or another Native American artist. Write a paragraph describing how you think the Southwest landscape is reflected in the work.

Broadening Your Understanding

Find a book with photographs of Georgia O'Keeffe's paintings. Then imagine you have to describe them to people who cannot see. Write a description of the work that will help people who are blind understand O'Keeffe's work and its power.

19 THE TOWER OF PARIS

When one thinks of famous *structures* of the world, the Eiffel Tower comes to mind. This *breathtaking* *landmark* rises 984 feet above Paris. It offers a grand *panorama* of the city.

A visitor can take an elevator to any of the tower's three observation floors. From each *platform,* the camera bug can take pictures of the city. These photos will surely become treasured *souvenirs.* The visitor gets a splendid view of the Cathedral of Notre Dame, the Seine River, and the Louvre, one of the world's leading art museums. *Binoculars* are available for closeup viewing of these great sights.

The tower was designed by the man for whom it was named—Alexandre-Gustave Eiffel. It was built for an *exposition* in 1889. Nothing like it had ever been built before. It was constructed in less than a year at small cost. The Eiffel Tower remained the tallest building in the world for 41 years. (The Chrysler Building in New York topped it in 1930.)

There is a restaurant on the first floor of the tower. Here, the visitor can hear *expressions* of wonder in many languages. Tourists from all over the world make a point of visiting the tower.

For the tourist from any nation, the memories of visiting the Eiffel Tower will *endure* forever.

UNDERSTANDING THE STORY

 Circle the letter next to each correct statement.

1. The main purpose of this story is to
   a. list the many historical landmarks in Paris.
   b. describe probably the most famous landmark in Paris.
   c. tell of the struggles that Alexandre-Gustave Eiffel had in building the tower.

2. From this story, you can conclude that
   a. the Eiffel Tower is taller than the World Trade Center in New York City.
   b. Paris is the most popular city in Europe.
   c. the loss of the Eiffel Tower would be a serious blow to French tourism.

MAKE AN ALPHABETICAL LIST

>>>> *Here are the ten vocabulary words in this lesson. Write them in alphabetical order in the spaces below.*

exposition	platform	landmark	panorama	souvenirs
expressions	breathtaking	structures	endure	binoculars

1. _____
2. _____
3. _____
4. _____
5. _____

6. _____
7. _____
8. _____
9. _____
10. _____

WHAT DO THE WORDS MEAN?

>>>> *Following are some meanings, or definitions, for the ten vocabulary words in this lesson. Write the words next to their definitions.*

1. _____ a wide view of a surrounding region

2. _____ very exciting; thrilling

3. _____ sounds or actions that show some feelings

4. _____ a raised level surface

5. _____ things that are built, such as buildings or towers

6. _____ something familiar or easily seen

7. _____ glasses used to magnify faraway objects

8. _____ public show or exhibition

9. _____ things bought or kept for remembrance

10. _____ to last; to keep on

112

COMPLETE THE SENTENCES

>>>> *Use the vocabulary words in this lesson to complete the following sentences. Use each word only once.*

souvenirs	binoculars	expressions	structures	platform
breathtaking	exposition	panorama	endure	landmark

1. At an _____, products of science, industry, and art are displayed.

2. Among the well-known _____ in Paris, the Eiffel Tower is the most famous.

3. Just as the Washington Monument is a national _____ in the United States, the Eiffel Tower is one in France.

4. The view from the observation deck is _____.

5. This deck serves as an ideal _____ from which to take pictures.

6. Tourists return with _____ from their tour of Europe.

7. Many tourists carry _____ with them so they can get close-up views.

8. You can hear _____ as visitors look at the view.

9. For a _____ of New York City, go to the top of the World Trade Center.

10. To help the Eiffel Tower _____ for years to come, it is well cared for.

USE YOUR OWN WORDS

>>>> *Look at the picture. What words come into your mind other than the ten vocabulary words used in this lesson? Write them on the lines below. To help you get started, here are two good words:*

1. _____ graceful _____
2. _____ tall _____
3. _____
4. _____
5. _____
6. _____
7. _____
8. _____
9. _____
10. _____

DESCRIBE THE NOUNS

>>>> *The two vocabulary words below are nouns. List as many words as you can that describe or tell something about the words* panorama *and* structure. *You can work on this with your classmates.*

panorama

1. _____
2. _____
3. _____
4. _____
5. _____
6. _____
7. _____
8. _____

structure

1. _____
2. _____
3. _____
4. _____
5. _____
6. _____
7. _____
8. _____

COMPLETE THE STORY

>>>> Here are the ten vocabulary words for this lesson:

breathtaking	landmark	souvenirs	panorama	exposition
platform	expressions	binoculars	endure	structures

>>>> *There are six blank spaces in the story below. Four vocabulary words have already been used in the story. They are underlined. Use the other six words to fill in the blanks.*

When the Eiffel Tower was built for an _____ back in 1889, it was the tallest of all <u>structures</u> in Paris. Today this _____ is still one of the wonders of the world.

Many _____ of wonder have been used to describe the Eiffel Tower's view. Indeed, the observation deck offers a <u>breathtaking</u> _____ of the city. A visitor can take a close look by using _____.

Tourists may take advantage of a restaurant on the bottom <u>platform</u> and a stand where they may purchase _____.

The Eiffel Tower will continue to <u>endure</u> as one of the world's premier attractions.

Learn More About Architecture

>>>> *On a separate sheet of paper or in your notebook or journal, complete one or more of the activities below.*

Working Together

Design a city of the future with your classmates. Divide up responsibilities so that different people are responsible for different parts of the city design. Parts of the city that need to be designed include streets, office buildings, stores, homes, recreational buildings, schools, hospitals, and public buildings, such as libraries and government offices. When everyone has designed his or her part of the city, draw a map. The map should show the city's buildings as if someone were looking down on the city from an airplane. You may also want to include drawings that show what the city will look like at street level.

Learning Across the Curriculum

Today skyscrapers are constructed in ways that the designer of the Eiffel Tower wouldn't have dreamed of. Find out some details about how skyscrapers are made today and write a brief explanation of what goes into the construction. If you wish, make a diagram of the inside of a skyscraper and explain some of the elements of its design.

Broadening Your Understanding

Check out a guidebook of Paris. Read about the architecture of the city. Then write a description of the five buildings you would most like to see in Paris and why. Explain something about the history of each building.

On a *humid* summer night in 1986, Midori Goto prepared to take a bow. The 14-year-old violinist was *stunned* by what happened next. The audience cheered and whistled wildly. Then the *conductor* and musicians of the Boston Symphony *Orchestra* hugged and kissed her.

The young concert violinist had just played Leonard Bernstein's *Serenade.* Her performance was "technically near-perfect." But that was only part of the story. Near the end of this long, *elaborate* piece, one of the strings on her violin snapped. Without missing a beat, she borrowed the concertmaster's violin and continued. Then, unbelievably, another string broke. Again she switched violins and continued playing. The new violins were larger than the one this *diminutive* girl was used to. But Midori finished the concert. She later explained, "I didn't want to stop. I love that piece."

Japanese-born Midori Goto was a child *prodigy.* At age 10, Goto was accepted by the *renowned* Juilliard School in New York City. She went on to play with some of the world's best-known violinists.

But on that magical night in 1986, Goto's whole life changed. She became famous. Yet Goto remained an *unaffected* teenager. While on tour, she still liked to *frolic* with her Snoopy doll.

Now in her twenties, Goto gives free concerts for children in the United States and Japan. She wants to share her love of music with them and get them as excited as she is about music from the great composers. Sometimes, she even gives free violin lessons!

UNDERSTANDING THE STORY

 Circle the letter next to each correct statement.

1. The main idea of this story is that
    a. Midori Goto switched violins twice during a performance.
    b. Goto has performed with some of the best-known violinists in the world.
    c. Goto has proved to be a talented violinist and a determined young lady who enjoys life.

2. From this story, you can conclude that
    a. Midori Goto will not give up easily.
    b. Goto will continue to break violin strings during her concerts.
    c. Goto will return to the Juilliard School.

MAKE AN ALPHABETICAL LIST

>>>> *Here are the ten vocabulary words in this lesson. Write them in alphabetical order in the spaces below.*

elaborate	frolic	orchestra	unaffected	diminutive
conductor	stunned	prodigy	humid	renowned

1. _____
2. _____
3. _____
4. _____
5. _____

6. _____
7. _____
8. _____
9. _____
10. _____

WHAT DO THE WORDS MEAN?

>>>> *Following are some meanings, or definitions, for the ten vocabulary words in this lesson. Write the words next to their definitions.*

1. _____ the leader of a group of musicians

2. _____ very small; tiny

3. _____ not influenced or changed; natural

4. _____ musicians who perform together, especially for playing symphonies

5. _____ having a great reputation; famous

6. _____ complicated; intricate

7. _____ damp, moist air

8. _____ a highly gifted or talented person, usually a child

9. _____ made senseless, dizzy; confused

10. _____ to play about happily

COMPLETE THE SENTENCES

>>>> *Use the vocabulary words in this lesson to complete the following sentences. Use each word only once.*

elaborate	unaffected	renowned	frolic	prodigy
humid	stunned	diminutive	orchestra	conductor

1. We don't expect to see famous musicians _____ with a doll.

2. It is difficult to perform on a hot, _____ night.

3. Teachers at the Juilliard School knew Goto was a _____.

4. Goto was _____ when the other musicians hugged and kissed her.

5. Goto, like all other musicians, had to follow the lead of the _____.

6. Her performance at Tanglewood proved that Goto deserved to be _____.

7. Goto looked especially _____ next to the taller musicians.

8. A musician must practice often to play an _____ piece like *Serenade*.

9. Goto's friends are glad that, in spite of her fame, she is still _____.

10. Members of the _____ were proud to be part of Goto's special night.

USE YOUR OWN WORDS

>>>> *Look at the picture. What words come into your mind other than the ten vocabulary words used in this lesson? Write them on the lines below. To help you get started, here are two good words:*

1. _____ performer _____
2. _____ violin _____
3. _____
4. _____
5. _____
6. _____
7. _____
8. _____
9. _____
10. _____

119

MAKE NEW WORDS FROM OLD

>>>> *Look at the vocabulary word orchestra. See how many words you can form by using the letters of this word. Write at least ten words in the spaces below.*

orchestra

1. _____
2. _____
3. _____
4. _____
5. _____
6. _____
7. _____
8. _____
9. _____
10. _____

COMPLETE THE STORY

>>>> Here are the ten vocabulary words for this lesson:

elaborate	unaffected	orchestra	frolic	renowned
humid	conductor	diminutive	stunned	prodigy

>>>> *There are six blank spaces in the story below. Four vocabulary words have already been used in the story. They are underlined. Use the other six words to fill in the blanks.*

Goto's parents knew their child was a prodigy. Yet even they must have been
_____ by her performance at Tanglewood. The conductor hugged and
kissed her. So did the musicians in the _____. They all realized that
the hot, _____ weather had made it hard to perform. They also
realized that the music was long and _____. How difficult this
experience must have been for the diminutive girl! Goto surely deserves to be
_____. But she never felt too important to _____
like a child. Who could have expected her to remain so unaffected after all that
attention?

Learn More About Musicians

>>>> *On a separate sheet of paper or in your notebook or journal, complete one or more of the activities below.*

Appreciating Diversity

Research your favorite musician. Find out what you can about the musician and write a report about him or her. What has helped the person succeed? What kind of music does he or she play?

Learning Across the Curriculum

The many kinds of instruments in an orchestra make music in different ways. Choose an instrument in which you are interested and find out the science behind the music. Write a description of how the instrument you chose makes music. You may want to make an illustration to help explain what you learn.

Broadening Your Understanding

Go to the library and check out one of Midori Goto recordings. Several of her recordings are listed below. Listen to her perform. Then write a review of what you hear. What parts did you like? Would you recommend a classical concert to your friends? If you cannot find one of Goto's recordings, listen to a recording by another violinist.

Bartok, *Violin Concertos 1 and 2*
Paganini, Niccolo, *24 caprices, op. 1*
Live at Carnegie Hall

Glossary

A

abrupt *[uh BRUHPT]* sudden

activist *[AK tiv ist]* a person who publicly supports a cause

actual *[AHK chu uhl]* real

affiliate *[af FIL ee it]* a person or an organization usually connected to a larger organization

altered *[AWL turd]* changed

anguish *[AYNG gwihsh]* pain; sorrow

apology *[uh PAWL uh jee]* an expression of regret for wrongdoing

apparatus *[ap uh RAT us]* materials, tools, special instruments, or machinery needed to carry out a purpose

apparel *[uh PAIR ul]* clothing; dress

apply *[uh PLY]* to seek a job; to ask for work

appropriate *[uh PRO pree it]* proper

approximately *[uh PROK suh mit lee]* nearly; about

aptitude *[AP tih tood]* a natural ability or capacity; a talent

arrested *[uh REST ed]* held by the police

assignments *[uh SYN mihnts]* definite tasks or jobs to be done; specific works to be accomplished

astronomical *[as truh NOM uh kul]* having to do with astronomy; the study of planets, stars, and other bodies in outer space

attained *[uh TAYND]* reached; achieved

auctioneer *[AWK shun eer]* the agent in charge of selling at an auction

authorities *[uh THOR ih teez]* specialists

await *[uh WAYT]* to wait for; to expect

B

banished *[BAN isht]* forced to leave one's country

banquets *[BAN kwits]* formal meals; lavish feasts

barriers *[BAIR ee uhrz]* obstacles; walls

binoculars *[buh NOK yuh lurz]* glasses used to magnify faraway objects

blessed *[BLES id]* given great happiness

bolted *[BOHL tid]* fastened; held with metal fittings

breathtaking *[BRETH tayk ing]* very exciting; thrilling

burial *[BEHR ee uhl]* having to do with placing a body in its final resting place

C

carnage *[KAHR nij]* the killing of a great number of people or animals

cello *[CHEL oh]* a musical instrument similar to a violin, but much larger, that is played in a sitting position

challenge *[CHAL unj]* a call to a contest or battle

chant *[CHANT]* rapid and rhythmic speaking

chasms *[KAZ umz]* deep openings or cracks

circumstances *[SER kuhm stans ehz]* conditions

clamor *[KLAM ur]* to demand noisily; to call for loudly

classification *[klas ih fih KAY shun]* putting something into a special group or class

commend *[kuh MEND]* to praise; to acclaim as worthy of notice

compete *[kom PEET]* to oppose; to try for the same thing

conductor *[kuhn DUK tuhr]* the leader of a group of musicians

conservationists *[kon sur VAY shu nists]* persons who wish to save forms of animals and plant life in danger of being destroyed forever

considerable *[kun SID ur uh bul]* not a little; much

contract *[KON trakt]* a legal paper promising a job

cradle *[KRAY dul]* a framework upon which a ship rests, usually during repair

cremated *[KREE mayt id]* burned to ashes

crescent *[KRES unt]* the shape of the moon in the first or last quarter; the symbol of the Muslim religion

cultural *[KAHL chur uhl]* relating to the beliefs and behaviors of a social, ethnic, or religious group

D

dealt *[DELT]* handled; managed; faced

debut *[DAY byoo]* a first appearance before the public

define *[duh FEYEN]* to explain the meaning of

definitely *[DEF ih nit lee]* absolutely

denote *[dih NOHT]* to show; to point out

despite *[dih SPEYET]* not prevented by; in spite of

devoted *[dee VOH tid]* gave up one's time, money, or efforts for some cause or person

dictatorship *[dik TAY tur ship]* the rule by one person or group that must be obeyed; power held by a few through force

diminutive *[dih MIN yoo tiv]* very small; tiny

disciplinarian *[dis uh plih NEHR ee uhn]* a person who believes in strict training

documentary *[dok yuh MEN tuh ree]* a factual presentation of a scene, place, or condition of life in writing or on film

dwindled *[DWIN duld]* reduced in number

dynamic *[deye NAM ihk]* full of energy; vigorous

E

eclipses *[ee KLIP siz]* times when the sun or moon cannot be seen because its light is blocked

efforts *[EF furts]* attempts

elaborate *[ih LAB or uht]* complicated; intricate

embodies *[em BOD eez]* represents in real or definite form

endure *[en DYOOR]* to last; to keep on

enforce *[en FORS]* to make someone do something; to compel

enthusiastic *[en THOO zee AS tik]* eagerly interested

erupt *[ee RUPT]* to explode; to burst forth

essential *[ee SEN chul]* absolutely necessary

exert *[eg ZURT]* to apply; to use fully

exhibited *[eks ZIH bih tid]* displayed; shown to the public

expedition *[ek spuh DISH un]* group of people undertaking a special journey, such as mountain climbing

explanation *[ek spluh NAY shun]* a statement that clears up a difficulty or a mistake

exposition *[ek spoh ZISH un]* public show or exhibition

expressions *[ek SPRESH unz]* sounds or actions that show some feelings

extermination *[ek STUR mih NAY shun]* the act of destroying completely; putting an end to

F

falter *[FAWL tuhr]* to hesitate; to fail or weaken

fan *[FAN]* an enthusiastic supporter

fare *[FAIR]* the cost of a ticket

fascinated *[FAS ih nay tid]* amazed; very interested by

festivals *[FES tih vulz]* celebrations

fiction *[FIK shun]* a story that is not true

flawless *[FLAW les]* perfect; without fault

flirted *[FLER tihd]* showed an interest

forbears *[FOR behrz]* family members who lived a long time ago

frantic *[FRAN tik]* wild with excitement; out of control

frolic *[FRO lik]* to play about happily

G

gingerly *[JIN jur lee]* very carefully

gloss *[GLOS]* high polish; shine

grave *[GRAYV]* serious; critical

grudge *[GRUHJ]* resentment; ill feelings

H

hardy *[HAHR dee]* able to take hard physical treatment; bold; daring

harpoon *[hahr POON]* a long spear with a rope tied to it used in killing a whale

hoax *[HOKS]* a trick

humid *[HYOO mid]* damp, moist air

I

identify *[eye DEN tuh fy]* to recognize as being a particular person or thing

ignorance *[IG nur uns]* a lack of knowledge

impressionistic *[im preh shun IS tik]* in the style of painting in which the painter tries to catch a momentary glimpse of the subject

informally *[ihn FOOR mal lee]* a way of doing something that does not follow exact rules or procedures; casually

instructors *[ihn STRUHK tuhrs]* teachers; leaders

intensive *[ihn TENS ihv]* concentrated

interior *[in TEER ee ur]* inside; inner part

investigate *[in VES tih gayt]* to study; to look into carefully

issues *[ISH yooz]* topics or problems under discussion

J

jointly *[JOINT lee]* together; in partnership

L

laden *[LAY din]* loaded; heavily burdened

landmark *[LAND mark]* something familiar or easily seen

lava *[LAH vuh]* melted rock that comes from a volcano

literacy *[LIT ur uh see]* the ability to read and write

M

mammals *[MAM ulz]* animals that feed milk to their young; people belong to this group

manager *[MAN ij ur]* a performer's business arranger

margin *[MAHR juhn]* a border; the space allowed for something

matter *[MAT ur]* a real thing; content rather than manner or style

mere *[MIR]* only; barely

mingle *[MING gul]* to mix; to get along together

misery *[MIZ uhr ee]* suffering; distress

mission *[MIH shun]* a special task

monarch *[MON ahrk]* a king or queen; an absolute ruler

monument *[MON yuh mehnt]* something from a past age that is believed to have historical importance

motivated *[MOH tuh vayt id]* stimulated to do something; inspired

N

notable *[NOH tuh bul]* worthy of notice; remarkable

novels *[NAWV uhlz]* long stories about imaginary people and events

O

obligations *[awb luh GAY shuns]* duties; responsibilities

orchestra *[OR kes truh]* musicians who perform together, especially for playing symphonies

ornamented *[OR nuh men tid]* decorated; made more beautiful

ovations *[oh VAY shunz]* bursts of loud clapping or cheering; waves of applause

P

pagan *[PAY gun]* a follower of a religion with many gods

panorama *[pan uh RAM uh]* a wide view of a surrounding region

peak *[PEEK]* the top; the highest point

pedigree *[PED ih gree]* the record of an animal's ancestors, especially with respect to purity of breed

perilous *[PEHR uh lus]* dangerous; hazardous

perished *[PEHR isht]* died, usually in a violent manner

permanent *[PUR muh nehnt]* lasting; continuing

perplex *[pur PLEKS]* to puzzle; to confuse

persisted *[puhr SIHST uhd]* continued in spite of obstacles

plastic *[PLAS tik]* a synthetic or processed material

platform *[PLAT form]* a raised level surface

plight *[PLYT]* a condition or state, usually bad

predicted *[pree DIHKT uhd]* described what would happen in the future; forecasted

preferable *[PREH fir uh bul]* something liked better; more desirable

prejudice *[PREJ uh dihs]* dislike of people who are different

primitive *[PRIM uh tiv]* living long ago; from earliest times

probable *[PRAWB uh buhl]* likely to happen

prodigy *[PROD uh jee]* a highly gifted or talented person, usually a child

products *[PROD ukts]* manufactured items

profession *[pruh FESH un]* an occupation requiring an education

project *[PRAH jekt]* an undertaking, often a big, complicated job

proof *[PROOF]* facts; evidence

prosperous *[PROS per us]* successful

protest *[PROH test]* strong objection; opposition

publication *[pub le KAY shun]* the production of written material into printed form

pundits *[PUN ditz]* persons who have knowledge of a subject

pursuing *[pur SOO ing]* striving for

R

racial *[RAY shul]* of or having to do with race or origins

recognition *[rek ugh NIHSH uhn]* special notice or attention

recognized *[REK uhg NYZD]* identified

reflect *[ree FLEKT]* to give back an image of

religious *[ree LIJ us]* having to do with a belief in God; devout

renditions *[ren DISH uns]* performances or interpretations

renowned *[rih NOWND]* having a great reputation; famous

reputable *[REP yuh tuh bul]* honorable; well thought of

resembles *[ree ZEM bulz]* looks like; is similar in appearance

restless *[REHST luhs]* uneasy; bored

restored *[rih STORD]* brought back to its original state; reconstructed

restricted *[ree STRIKT id]* limited in freedom or use

retrospective *[RET roh SPEK TIV]* an exhibition of the life work of an artist

romanticist *[roh MAN tih sist]* one who paints people and things as she or he would like them to be rather than as they really are

rural *[RUR ul]* having to do with open country and farming

S

sacred *[SAY krid]* holy; worthy of reverence

salvage *[SAL vij]* the act of saving a ship or its cargo from the sea

scale *[SKAYL]* to climb

scholarship *[SKOL ur ship]* money given to help a student pay for studies

segregation *[seg ruh GAY shun]* separation from others; setting individuals or groups apart from society

sheer *[SHEER]* steep; straight up and down

shrines *[SHRYNZ]* sacred places; places where holy things are kept

shy *[SHY]* modest; uncertain

sire *[SEYER]* to be the father of

skeptical *[SKEP tuh kul]* having doubts; not willing to believe

skillful *[SKIL ful]* having ability gained by practice or knowledge; expert

smart *[SMAHRT]* to feel a sharp pain; to sting

soared *[SOHRD]* rose upward quickly

souvenirs *[SOO vuh NEERZ]* things bought or kept for remembrance

spectacular *[spek TAK yuh lur]* eye catching; very unusual

structures *[STRUK churz]* things that are built, such as buildings or towers

stunned *[STUND]* made senseless, dizzy; confused

sullen *[SUL uhn]* gloomy; resentful

summit *[SUM it]* the peak; the highest point

symbols *[SIM bulz]* things that stand for or represent something else; signs

synonymous *[si NON uh mus]* alike in meaning or significance

T

talent *[TAL unt]* a natural gift for doing something

trade *[TRAYD]* a job; a skill

tremendous *[tru MEN dus]* huge; enormous

tyranny *[TIR uh nee]* the cruel use of power

U

unaffected *[UN uh FEK tid]* not influenced or changed; natural

urged *[URJD]* advised strongly

V

vanish *[VAN ish]* to disappear

vast *[VAST]* huge; spacious

vigor *[VIHG uhr]* strength; vitality

volcano *[vol KAY noh]* a mountain with a cuplike crater that throws out hot melted rock and steam

vulnerable *[VUL nur uh bul]* defenseless against; open to attack or injury

W

welfare *[WEL fair]* happiness; well being